Military Organization
and Society

MILITARY ORGANIZATION
AND SOCIETY

by

STANISLAV ANDRESKI

with a foreword by
A. R. Radcliffe-Brown

UNIVERSITY OF CALIFORNIA PRESS
Berkeley and Los Angeles 1968

Foreword

BY PROFESSOR A. R. RADCLIFFE-BROWN, F.B.A.

THE name 'sociology' is nowadays applied to a great many different kinds of writing about society, but when it was invented by Auguste Comte he intended it to designate a positive, inductive science of human society. Such science may be said to have had its first significant development in the work of Montesquieu in the middle of the eighteenth century. The author of the present book belongs to the sociological tradition that, starting from Montesquieu, includes such thinkers as Herbert Spencer, Émile Durkheim and Max Weber. The idea formulated by Montesquieu is that there are important relations of interdependence amongst the various features of social life that characterize different societies, and he applied this idea in an attempt to discover the relations between the laws of society and other features of social life, the form of government, the religion, the economic institutions, usages of various kinds and geographical environment. Such relations of interdependence or, to use Comte's term, 'consensus' can only be discovered by a comparative study of many societies of diverse types. It is for this reason that Herbert Spencer referred to the subject on which he wrote as 'comparative sociology'. The scientific use of this comparative method is admirably illustrated in this most remarkable book by Dr. Andrzejewski. He employs it not in order to construct a grandiose sociological system, but to deal with a properly limited though nevertheless extremely intricate and interesting set of problems.

The characteristic of science since Galileo is the use of experimental method. This term is frequently taken to denote experimentation in the sense of operations by which an event to be observed is brought about by the experimenter. But the Latin 'experiri' means only 'to put to test'. What the experimental

v

method, in the broad sense of the word, really is, is a method of investigation and reasoning in which general ideas are systematically tested by reference to carefully observed facts. As Claude Bernard says in his *Introduction à l'étude de la médicine expérimentale*: 'The experimental method, considered in itself, is nothing other than a reasoning in the aid of which we methodically submit our ideas to the test of facts. The reasoning is always the same, in the sciences which study living beings just as much as in those which are concerned with inanimate bodies. But, in each kind of science, the phenomena vary and present a complexity of their own.' In social sciences the only way of applying the experimental mode of reasoning is by comparing diverse forms of social life and their changes.

Observation is only scientific when it is directed by some hypothesis, some tentative generalization. Although it is true that fruitful hypotheses can be conceived only on the basis of a survey of facts. Charles Darwin wrote: 'How odd it is that anyone should not see that all observation must be for or against some view if it is to be of any service.' Claude Bernard, again, wrote: 'The experimental method cannot give new and fruitful ideas to men who have none; it can serve only to guide the ideas of men who have them, to direct their ideas and develop them so as to get the best possible results. As only what has been sown in the ground will ever grow in it, so nothing will be developed by the experimental method except the ideas submitted to it. The method itself gives birth to nothing.' The most difficult and important task in science is that of formulating problems. It can even be said that it is much harder to know what kind of questions to ask than it is to answer them. It is from this point of view that the value of the present book ought to be judged. There can be no doubt that it contains an extraordinarily large number of extremely penetrating and original ideas of fundamental importance.

The formation and development of an inductive science requires two things. One is the provision of some acceptable systematic classification of the phenomena with which the science has to deal; the other is the creation of a coherent set of technical terms; these two tasks are interconnected. The value of classifications and terminologies lies in their essential service in the discovery of valid generalizations. As Whewell aptly puts it: 'The fundamental principle and supreme rule of all scientific terminology is that terms must be constructed and appropriated so as to be fitted to enunciate simply and clearly true general propositions.' These considerations should be borne in mind by those who may feel inclined to object to the terms proposed by Dr. Andrzejewski. The provision of heuristically useful classifications and terminologies is a condition of further progress of social sciences, but this task is so difficult that serious, let alone successful, attempts in that direction are quite exceptional.

Comparative sociology presents considerable difficulties to its students, whether they are the writers of books or the readers. This is so because in any attempt to arrive at conclusions that will be valid for society in general it is necessary to take into account a large number of sufficiently diverse forms of social life and to deal with a large body of factual data derived from historical and sociographic literature. The immense labour involved in acquiring adequate range of information is undoubtedly one of the main reasons why so few writers on social sciences undertake investigations of this kind. Therefore, every attempt in this field deserves to be welcomed.

The slow progress of comparative sociology as an inductive science dealing with the whole range of the social life of human beings is not entirely due to the difficulties of the study itself.

There is amongst social scientists a preference for facts over theories in the belief that only factual knowledge is of immediate utility in practical life. They ignore the fact well known to physical scientists that it is purely theoretical investigations which ultimately lead to the most important practical results, as when Clark Maxwell's effort to find a better formulation of the equations expressing Ohm's and Faraday's Laws led to the hypothesis of the existence of hitherto undiscovered electro-magnetic waves, and so to the inventions of such things as radio and radar. The great advances of applied science and technology of the past three and a half centuries could not have taken place without the devotion of scientists to purely theoretical enquiries. And there can be no doubt whatsoever that it is the wider application and the refinement of the method of comparative sociology—the investigations such as this—that hold out the promise of a really scientific understanding of human society.

A. R. RADCLIFFE-BROWN

Preface

THIS book grew beyond what it was first intended to be. Owing to this expansion the character of the exposition varies in different parts of the book: the empirical evidence is described much more fully in the earlier parts of the book than in the later. Nevertheless, the present arrangement has this important advantage that it introduces the reader gradually into the bewildering complexity of the phenomena under analysis: at the beginning the theoretical argument is fairly simple, and only gradually it becomes concentrated abstract and involved. The present arrangement can also be defended on the ground that only by expanding the book tenfold could I make the exposition of the theories as diluted in the latter parts as it is in the first part of the book. All theoretical comparative studies face this dilemma: either they attempt to include detailed descriptions of the factual data on which the conclusions are based, and so acquire the forbidding bulk of Frazer's Golden Bough which drowns the theoretical framework; or they merely mention the instances, and so become rather ethereal. If the ideas advanced are complex the whole becomes somewhat difficult to assimilate. I do not maintain that I have altogether avoided the latter blemish, but I think that I am justified in pleading that the problems I dealt with cannot be discussed in a way which is suitable for bedtime reading. Perhaps I have erred on the side of excessive brevity, but my consolation is that no reader will have to wade through a morass of empty verbiage or irelevant details in search of theoretical ideas. In order to forestall possible accusations of superficiality I must say that, when describing an instance bearing upon my theory, I tried to put down only what is strictly relevant to the argument. I did not consider it my duty to say everything I

ix

knew about a case. It must be remembered that I am not writing a history, where fullness of description is a merit, but expounding a theory.

In spite of Max Weber's and Durkheim's brilliant contributions, which demonstrated the proper way of studying social phenomena, their lead has not been followed. With all the heaping up of descriptive studies there has been a great dearth of ideas. In my opinion, most of the theories (though fortunately not all) propounded recently scarcely amount to more than verbose and pompous refurbishings of simple truths known since the days of Aristotle. The works which have put forth new ideas are astonishingly few in relation to the spate of sociological literature which came out during the last thirty years. The reason for this *'misère de la sociologie'* is that it is very easy to spin out vague unverifiable theories, and it is not too difficult to produce a straightforward description of a concrete situation, but it is very difficult to construct a general theory which fits all the known facts, without being tautological. I think that the theories expounded on the pages which follow do explain a considerable number of facts, and I hope that this might induce the reader to forbear various blemishes of this work, of which I am only too painfully aware. Indeed, I had to force myself to abandon further attempts to improve it, or else it would never have been written. I came to the conclusion that whatever I do the exposition will be far from perfect. In any case, a pioneering monograph can hardly compete in this respect with manuals following well-trodden paths. I thought that it is more meritorious to put forth some new ideas, even in an imperfect form, than to add another item to the already impressive array of didactically perfect assemblages of pre-digested platitudes.

At the risk of incurring the wrath of purists, I invented a few terms. No science can develop without a special terminology. The lack of such a terminology in sociology and 'political science' is a proof that these sciences hardly exist, because it should be obvious to anybody who has reflected upon this matter that ordinary words are quite inadequate for describing social structures. Furthermore, simple sentences containing new terms may be easier to understand than long phrases employing only well-known words. For these reasons a number of neologisms will be found on the pages which follow. In a way this makes the reading more difficult. But I do not think that it is unreasonable, even though it may be unusual, to expect from a reader of a book on sociology, the modicum of mental effort required for assimilating a very elementary book on, say, logic or physiology. Naturally, a terminology may be good or bad—that is to say, helpful or misleading—but that can be decided only by those who pursue further the same lines of enquiry.

I have almost dispensed with bibliographic footnotes. Some references I have inserted in the text. I see no reason why the reader should be condemned to search somewhere at the bottom of the page or at the back of the volume, or, as in some maliciously arranged books, in both places, for something that he could read in the text. These references generally pertain to theoretical argument. No bibliographic indications lending support to descriptive statements are given, because to do it adequately would make reading unbearable, as nearly every sentence would have to be provided with a footnote, often several books would need to be quoted in support of one sentence, and some titles would have to be repeated innumerable times. I thought that a much better arrangement would be to indicate in the bibliographic appendix the sources of my information.

This book was virtually completed in 1950, and therefore certain passages in the last chapter might seem out of date. Notwithstanding, I left them as they were because some of the predictions made there have already been fulfilled, and this is also a test of the value of a theory. Some of the ideas expressed in the last chapter have in the meantime been given a wide currency by the three recent books by Bertrand Russell. Although this diminishes the novelty of the present book, I am glad that this great philosopher tries to enlighten people about these simple, but nearly always forgotten, truths.

This book was written while I was on the staff of Rhodes University, South Africa, although the preparations for it reach back to my school days, when I was keenly interested in strategy and military history, which I later abandoned for sociology. During the war, I had the opportunity to observe many different armies, sometimes from rather unusual points of vantage, and to ponder over the problems of military organization.

STANISŁAW ANDRZEJEWSKI

London, August 1952

Preface to the Second Edition

IDEALLY, a book written almost twenty years ago ought to be rewritten, particularly when the author feels that the presentation could be improved. For various reasons, however, I have decided to reprint the old text and to incorporate amplifications into the appendices. Apart from the desire to keep the price as low as possible, the most personal reason for this choice was that a complete revision would require interruption of my current work, which I would regard as worth while only if I felt that the propositions formulated in the first edition were basically wrong. In fact these propositions need no substantive alterations and my only regret is that I have couched some of them in neologisms which I now regard as superfluous. I have already made a confession on the latter point in the first chapter of *Elements of Comparative Sociology*, where I have also formulated the rules which ought to govern admissibility of neologisms.

In one of the appendices the reader will find self-explanatory substitutes for those of the original neologisms which I now regard as defective.

On a few points of fact some qualifications or amplifications might usefully have been made. They have not, however, been inserted because (apart from the consideration of printing costs) they would not have made much difference to the theoretical argument. In any case a theoretical and comparative book of this scope can never become a really reliable source of factual data, which any sensible scholar will seek in narrowly focused monographs written with the aim of providing such information rather than constructing a theory.

Another reason for reprinting the old text, is that it contains several forecasts which have since come true, such as the rejection of the complacent assumption (practically universal in the West in the early fifties) that democracy guarantees a technological lead. Other examples to the point are the forecasts that a Sino-Soviet rift will save the West; that the latter (especially the United States) will become less democratic and libertarian, and that democracy will fail in underdeveloped countries including India. For uttering the first I was reprimanded by various experts in communist affairs, and dismissed as an unworldly theoretician, while the third brought upon me the accusation of being a fascist.[1] The point of mentioning this is that, whereas lucky guesses based on hunches can only provide a ground for private self-congratulation, forecasts which derive from specific and explicit views on social causation enhance the latter's credibility if they turn out to have been correct.

The few methodological pronouncements contained in the Preface and the Introduction to the first edition no longer represent exactly my present viewpoint; and though not basically wrong, they would have to be slightly qualified in the light of what is said in the first four chapters of *Elements of Comparative Sociology* (Weidenfeld & Nicolson, 1964; also published by the University of California Press under the title of *The Uses of Comparative Sociology*).

The appendices attached to the present edition constitute additions rather than corrections. Some remarks relevant to the subject of the present book can also be found in *The Uses of Comparative Sociology*. The general factors which lead to military rule in underdeveloped countries are analysed at considerable length in *Parasitism and Subversion: the Case of Latin America* (Weidenfeld & Nicolson, London, 1966, and Pantheon Books, New York, 1967) and in *Neocolonialism and Kleptocracy: Africa's Political, Economic and Cultural Predicament*, which will appear about the same time as this edition. It might be worth mentioning that the

[1] I have been similarly castigated for the view that foreign aid to underdeveloped countries remains condemned to futility unless it is accompanied by a rapid diffusion of family planning; particularly as I have made a suggestion (outrageous to many even now, let alone in the fifties) that aid ought to be made conditional upon the willingness of the receiving governments to adopt vigorous policies of birth control. While the demographers and the economists were making development plans based on the rates of population growth existing at the time (that is to say 1950), I made (as the reader will see) the correct prediction that without effective population policies the rates of growth will rise with every improvement in the living conditions of the depressed populations, thus nullifying the benefits. The populations of the underdeveloped countries would by now be growing even faster than they are doing, had the Malthusian positive check of internal violence not have come into operation with rapidly increasing force.

theories developed in the present book dovetail with the findings of the three books just mentioned, which ought to add to their individual as well as joint credibility.

Perhaps the most valid justification for reprinting the original text unchanged is that it has opened one of the most flourishing fields of sociological and politilogical inquiry, and younger scholars may be interested to know how it all originated. Those interested in the development of the method might also like to see the diagrammatical presentation, which was first to introduce into sociological theory the 'box-and-arrow' models (cybernetic *avant la lettre*) which have become very popular, though often misunderstood, in recent years. To be sure, as can be seen from the bibliographic appendix to the first edition, a number of historians and political philosophers (as well as some of the founders of sociology, beginning with Aristotle and ending with Max Weber) have made important remarks about the impact of military organization on society; but they did so incidentally while concentrating on other problems. Many works have, of course, been written about the psychology of the soldier, the causes of wars and proposals for peace, but not about the relationship between military organization and social structure. And this brings us to the main reason for republishing this book.

Despite the profusion of works on sociology of military organization, few of the problems raised in this book have been taken much further. Apart from the always abundant strategic studies, nearly all the existing literature deals with the issues broached in Chapters VIII and XI—that is to say, those of military incursions into politics. Up till now no other book has appeared which would attempt to offer a systematic and comparative study of the impact on social structure of such factors as changes in weapons, in methods of warfare or in the military participation ratio; although a number of case studies have appeared which provide a further confirmation of my theories on various specific points, and from which I would have drawn additional supporting data if I were rewriting the book. Though non-tautological and therefore perfectly testable, these theories have not been invalidated and, as the reader will see, can explain a number of social changes which have taken place since they were first formulated.

When the first edition appeared the post-war optimism about the spread of democracy had not yet subsided. Despite the Cold

PREFACE TO THE SECOND EDITION

War most people expected that at least in internal politics the armed forces had ceased to play the preponderant part. Unfortunately, these hopes have been shattered, and especially during the last few years we have witnessed a resurgence of militocracy—whether of purely pretorian or more idelogically oriented variant—with the consequence that military dictatorships are more common today than at any time in history. Even the biggest communist state is not exempt from a tendency in this direction; and the most obvious result of the Great Cultural Revolution has been a taking over by the army of various functions and powers which in older communist states belong to the civilian bureaucracy. Furthermore, a few days before these words were written the Greek Army made a 'cuartelazo' which was the first in Europe since the last war.

Though in a different way, the armed forces of the United States—together with their appendages such as C.I.A.—are steadily moving towards the position of political dominance. Thus, unfortunately, the problem of the relations between military organization and society has lost none of its relevance to the great issues of our time.

STANISLAV ANDRESKI

Reading, April 1967

Contents

xvii

CONTENTS

Introduction

THE problem of the influence of military organization on society has, on the whole, failed to attract the attention of social scientists. To be sure, much has been written about war, its alleged evil or beneficial effects, its causes and the possibilities of its abolition. But the only writers who appreciated the importance of military factors in shaping societies were Max Weber and Gaetano Mosca. This persistent neglect is due, I think, to the insidious utopianism which pervades sociological thinking. Military organization influences social structure mainly by determining the distribution of naked power or, to use another word, the ability to use violence. Now, most writers are rather peaceful by nature and brute force is a thing which they would like to see exorcised for ever. On the other hand, those who are chauvinistic or inclined to worship might, are averse to critical examination of the exercise of violence, as such examination might besmirch the halo of their idols. Moreover, the belief in progress, which reigned indisputably until recently, included the conviction that humanity was becoming more and more peaceful. No need was felt, therefore, to occupy oneself with organized violence—the quickly disappearing relic of the barbaric past.

It is no doubt a dismal science, this analysis of the impact of naked power; and the guess about the future is none too reassuring. But I must inform the reader that I am not a dispeptic and disgruntled individual, bent on proving that the world is going to wrack and ruin. On the contrary, I am generally of optimistic disposition. But my supreme goal is to find the truth, and I have tried to the best of my ability to uphold every true scientist's credo of facing the facts, whether they are pleasant or not. I am convinced that, far from being a useless pastime, such disinterested pursuit of truth is the best way in which a social scientist can serve the

I

welfare of humanity. Science, being ethically neutral, cannot indicate the ultimate goals of human endeavour. Sociology cannot prove that one ought to love one's neighbour, any more than optics can prove that a painting is beautiful, or physiology, that chicken is tastier than beef. Chemistry can teach us how to cure, but also how to poison. Psychology can teach us how to rehabilitate juvenile delinquents, but also how to mislead by clever propaganda. Science, in short, is no substitute for ethics: it can only indicate means, never ultimate ends. Nevertheless, social sciences have an essential contribution to make. So far, social reformers have resembled more sorcerers than modern physicians. Knowing neither the causes of a disease nor the effective remedies, they tried and still try to cure by incantation; often trying to exorcise the evil by denying its existence. Modern medical science came into existence only after discoveries, which at the time seemed completely remote and useless, were made by men motivated only by the disinterested search for truth. Similarly, social sciences may give to social reformers the knowledge of the means by which they can attain their ends. They can never eliminate evil intentions, but at least they may help to avoid disasters so often brought about by actions of sincere, well-meaning men. In order to perform this task social scientists must observe as strict objectivity as is humanly attainable. They must resist the temptation to preach and admonish, at least in their scientific writings, because if they succumb to this temptation, their analytical thinking will be impaired, facts distorted, and the purpose of adding something to the stock of knowledge defeated. It is my earnest hope that by studying objectively what is perhaps the most tragic aspect of social life, without glossing over unpleasant facts and without an axe to grind, I may have helped to bridle the monster.[1]

I have not attempted to deal with all sociological problems related to the phenomenon of war. Some of them, of very great

[1] In order to avoid misunderstanding I must emphasize that I do not maintain that all normative discussions of human affairs are useless, or that sociologists must have nothing to do with such discussions. But I do contend that a clear-cut distinction between evaluative and existential judgments must be maintained. On this point I subscribe to the views of Charles Stevenson expressed in his *Ethics and Language* (Yale 1946). Notwithstanding, in order to satisfy those who claim that a writer on human affairs ought to declare his credo, I can say that on matters of ethics and philosophy in general I adhere almost without exception to the views of Bertrand Russell, who was my spiritual guide ever since my early youth when, dismayed by Hegel and his heirs and on the brink of concluding that philosophy is bunk, I read his *Outline of Philosophy*.

interest and importance, are hardly even mentioned: the question of biological selectiveness of various kinds of warfare, for example.[1] The aim of the present work is the examination of relations between military organization and social structure. And even in this field my task is limited. I have dealt with concrete events not for their own sake, but only in so far as they throw light on generalizations which I sought to establish. That is why this book is so slender. Many volumes would be needed for an investigation of how various unique features of various concrete military institutions were connected with variegated peculiarities of different societies. Such a treatise might be very interesting and useful, but it would be beyond the field of sociology, or its special branch— political science, whose supreme task is the search for generalizations of universal validity.

Particularly galling is the lack of really scientific terminology, which could aid one to describe social phenomena precisely. The only field of social sciences which possesses such terminology— the study of kinship organization—is unfortunately remote from the present enquiry. Elsewhere we are condemned to use vague, emotionally charged words of everyday language, even in dealing with economic phenomena, which are within the purview of the supposedly most exact of social sciences. But in reality the exactitude of economic theory is rather spurious. Its theoretical framework is like a steel construction on a foundaton of sand. It is precise within its platonic universe of discourse, but when it comes to the analysis of real economic structures and their transformations, then the only vocabulary available is that of the newspapers. When it comes to questions of political organization the position is even worse. Most of what goes under the name of political science is no science at all but simply propaganda; or at best a discussion of the historical and philosophical background of one's preferences. The terms used are loaded with sentiment and completely nebulous. The term democratic, for instance, simply replaced the old-fashioned word good. Soon perhaps we shall speak of democratic and communist or fascist soups and steaks. True, there are some descriptive studies, which are certainly very useful; but if, enticed by the sound of the word science, one looks for a body of established, empirically verifiable generalizations, one looks in vain, or

[1] This point is discussed in P. Sorokin, *Contemporary Sociological Theories* (New York 1928).

3

almost in vain.[1] One cannot say that these remarks are altogether inapplicable to the field tilled by those who call themselves sociologists, although it cannot be denied that here the situation is definitely better. At least there is less venting of loves and hates in a scientifically sounding disguise, and the dictum that the task of the scientist is to describe, analyse and explain—not to praise or blame—is almost universally accepted. On the other hand, many of the pseudo-scientific terms juggled about by some sociologists are just pompous substitutes for perfectly serviceable ordinary words, only more obscure and vague, often used in order to dazzle unsuspecting laymen with fictitious knowledge. Nevertheless, here some empirically verified generalizations can be found, though not in the quantity one could wish. The accumulation of such generalizations cannot proceed fast so long as the energies of most sociologists, particularly in the country where they are most numerous, are devoted almost entirely to writing innumerable text-books, mostly insipid, and to purely descriptive studies, often descending to the level of rather unintelligent inventories.

The present work will have fulfilled its purpose if it succeeds in enticing some social scientists to the field of studies, which combines abstract reasoning with empirical comparative investigation —the only road along which a body of really scientific generalizations can be built: a body of generalizations about relations of particular variables, I must stress, not an all-embracing system purporting to explain everything. This is the road shown by Max Weber, by far the greatest thinker to whom the name of sociologist can be applied.

The method of the present enquiry is comparative. This does not imply, of course, the adherence to the concept of unilinear evolution or any other a priori explanatory hypothesis. It simply means that in the search for regularities in the relations between military organization and social structure, I resorted to the comparison of different societies. In fact, there is no other method of establishing valid sociological generalizations. Not being able, unlike natural scientists, to manipulate relevant factors, all we can do is to observe variations occurring spontaneously. The connec-

[1] Harold D. Lasswell has the enormous merit of trying to make 'political science' into a real science. It is, therefore, particularly unfortunate that his thought is crippled by the vague, pseudo-scientific jargon.

4

tion between two or more phenomena can be proved not to be accidental only if it obtains when all other circumstances are different. For this reason, the more distant, unrelated and different are the societies under observation, the more instructive is their analysis.

I have not tried to use statistical methods, even though they are applicable in principle, because the gain in exactitude would be rather spurious. In the field of social sciences, where units must be arbitrarily delimited, the coefficient of correlation is not necessarily the most decisive proof of causal or structural connection. If we find that two phenomena occur together in two societies completely different in every other respect, and without contact with one another, then this is more conclusive proof that there is a causal or structural connection between them, than if they were found to coexist in a number of related neighbouring societies.

The fact that every social phenomenon is determined by several factors at the same time is the source of great difficulties. That is why the great majority of sociological generalizations can only be statements of tendencies. They can only assert that a given circumstance tends to produce certain results; that is to say, that it will produce them if there are no stronger counteracting factors at work. We should think about the interaction of social forces and its results in terms of vectorial analysis. In order to test an explanatory hypothesis, therefore, it is necessary to select cases where the most powerful interfering factors are absent. The difficulty of making valid generalizations in the early stages of any science is probably mainly due to the fact that, as very little is known about anything, the isolation of factors is not easily achieved. The most important step in any investigation is to single out possibly relevant factors and ascertain the direction of their influence. There is, however, one consolation. A coherent theory, which explains all known cases which come within its purview, is not likely to be basically wrong, even though not every intermediate generalization has been verified. The chance that it may fit so many facts just by accident is very slight indeed.

My hope is that the present work will stimulate further enquiry. Some of my conclusions will, no doubt, be found false; others may perhaps provide foundations for further discoveries. And though not all the generalizations proposed on the following pages can be regarded as verified, they are all verifiable. They are not statements of epistemological approach but definite theorems expounding connections between observable variables.

Many of my ideas are drawn from old philosophers, and I am convinced that there is more truth in their theories than is generally admitted nowadays. The approach of modern social scientists to the old masters is too often either that of ancestor worship, practised chiefly by 'political scientists', or of complete disregard for their thought, prevalent among the sociologists, who, if they ever read these works and find there some flagrant errors, discard them altogether. Both attitudes are wrong. The real task is to sift the true from the false, to re-interpret, to modify and to test; paying attention not so much to normative evaluations as to factual explanatory theories. Anybody who tries that will find that social sciences are cumulative like the natural sciences, and that we ought to be grateful to the old masters from Aristotle and Han Fei-tzu to Karl Marx and Herbert Spencer, because errors can be rectified by lesser men but it takes a genius to conceive an original idea.

Any reader familiar with the writings of Max Weber will notice my tremendous indebtedness to him. It is evident from almost every page. But this does not mean that the present book is the exposition of his theories. On the contrary, I tried to carry on from where he left off. Significantly, I have almost never found any of his statements definitely wrong. At most they needed modifications; but usually they provided excellent starting points for further analysis. The true measure of a thinker's greatness is the number of avenues for further enquiry which he opened.

I

Omnipresence of Struggle

THE most general assumption, on which the whole theoretical framework of this study rests, is the recognition of the
fact that the struggle for wealth, power and prestige (for
ophelimities, i.e. desirable things of life, as they may be called) is
the constant feature of the life of humanity. This statement does
not mean that I approve of 'this sorry scheme of things entire', any
more than the recognition of the inevitability of death implies any
liking for it. Whether we like it or not the fact is that no society,
no group however small has ever been heard of where such a
struggle would be altogether absent. Even associations of saintly
ascetics are not quite immune from it. It is beyond any doubt that
this universal trait of the social life of humans is due to the in-
effaceable characteristics of human nature. There is nothing of
that kind within societies of insects, but plenty of it among the
monkeys. I am not denying the existence of love; but this, in its
greatest intensity, is generally confined to a small circle, be it a
family or a few intimate friends. And even in such groups quarrels
take place. Anyway, this point is of no consequence to the present
argument. There is, of course, the feeling of intense group solidarity which often impels individuals to sacrifice even their lives.
But, even if we disregard the fact that such solidarity does not preclude inner friction, this argument does not invalidate our assumption. On the contrary, it is an essential elaboration of it. Indeed,
human beings, being social animals, always align themselves into
factions when they want to fight; and the intensity of the feeling
of solidarity varies inversely with the intensity of the feeling of
hostility towards outsiders. This fact is so clearly and easily perceptible that I shall not dwell on it; it is also confirmed by the
results of statistico-psychological investigations.

We might speculate on the problem of whether this tendency is

7

related to the innate propensity for fighting, if this exists. In favour of the view that it does exist may be adduced the fact that little boys all over the world appear to be addicted to pugilism; even in environments where such behaviour is frowned upon. But the alternative explanation of this fact is equally plausible: namely, that scuffling is the only way in which the desire for power and glory can be satisfied on the level of childish mentality. It is also probable that the feeling of belonging to a group which appears to be indispensable to human happiness, does require some measure of antagonism to other groups. Nevertheless, even the assumption of innate pugnacity would not explain the existence of war because war means killing, and there is no evidence that there is in human beings an innate desire to kill their own kind; on the contrary, it seems that apart from a relatively few sadists most men dislike it. Usually, killing is done for the sake of other ends.

The idea that constant struggle is an unavoidable ingredient of social life is not new. It was known to Greek philosophers, to Ibn Khaldun and to many others; it was systematically expounded by Ludwik Gumplowicz seventy years ago. The admission of the omnipresence of struggle does not imply the denial of the existence of other socio-cultural processes. Only a doctrinaire monist would assert that there are no such things as solidarity, compassion, mystic union, artistic creativity and so on. The struggle, more-over, does not always need to be violent or to preclude any com-promise and adjustment of interests. The irreducible residuum of it is the constant tendency of groups and individuals to acquire as much wealth, power and prestige as they can; and as these ophelimities are limited, struggle must result. But forms of struggle can be regulated: certain rules of the game may be imposed and the outcome decided without resort to the *ultima ratio* of naked violence. This is what happens in every society: everywhere a multitude of struggles is going on, in which violence is not used, such as business competition, missionary proselytizing and so on. Nevertheless, the use or the threat of violence is a very usual weapon.

Various psychologists, particularly those of the psychoanalytic persuasion, have spun ingenious, and often far-fetched theories, purporting to explain why men fight. But in my opinion we should turn the question and try to explain how it happens that men do not fight more often than they do. After all, it is quite in accordance with men's animal nature to seize the objects of their desires, by violence if necessary. Anybody who had to bring up children knows that they have to be taught not to fight about food, toys and other things. In this respect men are like other animals; and even their habit of fighting in groups has its parallels among their lowly relatives. I often watched dogs fighting, and I noticed that sometimes when several dogs were present they divided themselves into opposing alliances, just as men do in pubs and dance-halls. When thinking about sources of pugnacity we must always remember that very seldom men fight for the sake of fighting; usually they fight for something: be it food or women or precedence or what not. What goes under the name of pugnacity is rather a complex of tendencies than a single tendency. There is, I think, such a thing as pure pugnacity, but it is uncommon, and it arises through the mental process whereby a mode of behaviour, frequently adopted as a means, becomes by the force of habit a goal pursued for its own sake. The displaced aggressiveness—the 'taking it out on somebody else'—does not come under this heading because its aim is usually self-assertion or the humiliation of others. Sadism, i.e. the desire to inflict pain, should not be confused, as it is often done, with pugnacity, i.e. the desire to fight. On the basis of my observation of soldiers, policemen and gaolers of several different nations, I came to the conclusion that combative ardour and sadism are by no means so closely connected as the psychoanalysis would have us believe; to a large extent they even seem to be the opposites.

The animal nature of man accounts for the existence of fighting, though it does not explain why the latter assumes such variegated forms. The natural propensities of men do not account, however, for the systematic killing in which mankind indulges, because this practice is at variance with what goes on among other mammals. One of the chief reasons for this difference is the obvious circumstance, which is seldom taken into account in the discussions of this problem: namely, the fact that human beings use weapons. Fisticuffs usually end with thrashings or the flight of the beaten; killing cannot be a by-product of fighting, but it has to follow it and be done on purpose. Furthermore, in such a situation a victor does not have to fear the vengeance of the defeated. But it is not so if weapons are used, because then he who stabs or shoots first wins. Under such circumstances it is safest to kill one's enemies. Anyway, in all fighting where weapons are used some of the participants are likely to get killed. So we are justified in saying that the prevalence of killing within our species is the consequence of the acquisition of culture. This explains why *homo faber* might be named with equal justification *homo interfector*. The invention of weapons, however, had also some other even more far-reaching consequences about which I shall speak in a moment.

I have said that the struggle which is going on constantly within societies and between them is waged for power, wealth and prestige. All three are interwoven. Power is not necessarily coveted because of an innate desire for it, and the same can be said about wealth. Power may be desired because it leads to wealth, wealth because it gives power, and both because they bring prestige, which may be valued as a means of getting the other two. Also, in the final analysis wealth is a form of power: the power to dispose of things. Power and wealth are desired in the first place because they enable their possessors to satisfy their bodily needs. But power (and wealth, as one of its forms) may be desired for its own sake too. This desire is rooted, I believe, in an innate disposition of human beings, though it can be canalized, diverted, inhibited and repressed by social norms. It seems very probable that the strength of this propensity varies according to individuals, and probably sex. Much of this desire is, no doubt, due to the prestige-bringing properties of power. It follows that even if it is rooted in an innate disposition, it can be greatly weakened by social norms deprecating power, as has apparently happened among the Pueblo Indians. But there can be no doubt that the desire for prestige, the wish to be well thought of by others, is innate. It is a psychological element irreducible to any other, which, just as the innate sociability of human beings, is a precondition of all society and culture. In the last analysis, therefore, the objects of human strife can be resolved into the means of satisfying bodily needs and the means to satisfy the craving for prestige, power for its own sake being the possible third. In consequence, we can safely predict that some kind of struggle will always go on in human societies because prestige, as well as power, is relative and, therefore, there can never be enough of it for everybody. But the struggle for the sake of prestige alone can be regulated by social norms; it has been and is being so regulated in innumerable societies; and there is nothing that makes the extension of such regulation to the whole world inherently impossible. Such struggle can be carried on without killing. Killing one another remained the constant occupation of men decause they were obliged to fight for the means to satisfy their elementary bodily needs—chiefly food.

One of the greatest illusions is the notion that wars are always fought for illusions. Naturally, the hopes which motivate an attack or defence may prove vain. But that does not mean that the stake was illusory any more than the failure of a businessman's plan proves that he was pursuing unreal aims. This illusion is particularly widespread among British and American social scientists. Psychologists and ethnologists are busy finding out that it was the manner in which Japanese mothers-in-law treat their daughters-in-law or the Oedipus complex of German men that were responsible for the last war. It never occurs to these investigators that these peoples might have been fighting for very real stakes, and that had they won they would be much better off economically than they ever were. A Japanese peasant who hoped to obtain an extensive land-holding in Manchuria was won to the cause of war by a very real attraction. So was a German factory hand, perhaps unemployed until Hitler came to power, who was enchanted at the prospect of becoming an overseer in a factory in Russia. It does not follow that the last war was one of 'have-nots' against 'haves'. The fact that Japan attacked China, Germany—Poland, and Italy —Greece, precludes such interpretation. In these cases those who felt they did not have enough decided to get more by robbing those who had even less.

I by no means maintain that all wars are motivated by solidary interests of whole nations. Equally well they may be provoked by sectional interests. They may be instigated by merchants wanting markets, or officials wanting new posts, or scions of the gentry wanting estates, or capitalists wanting fields for investment. The mere lust for power and glory on the part of rulers is of enormous consequence too. But what demands explanation is the fact that these sectional interests were always able to find plenty of men willing to fight for them. This can only be explained on the ground that nearly all countries were full of desperadoes, ready to risk anything in order to escape misery or the fall from the customary standard of living. 'The ambition of princes would want instruments of destruction if the distress of the lower classes did not drive them under their standards,' says Malthus (*Essay*, p. 400).

Undoubtedly, some peoples are more warlike than others. But these differences cannot depend on the degree of severity of the methods of upbringing, as some psychologists would have us believe, because in some of the most warlike societies children are treated with the utmost laxity; among the Nyars or the Plains Indians, for instance. The family pattern of the Eurasian nomads was much less patriarchal than that of the Chinese, and nevertheless it was the nomads who were the perennial aggressors. The long-suffering Hindu peasants live in patriarchal families where women and children are mere chattels of the patriarch. The patriarchal and disciplinarian streak in the German family pattern is blamed for the aggressiveness of the German state. But the German family was no more egalitarian during the three centuries preceding 1850, when Germany was a victim of foreign invasions. When Commodore Perry attacked Japan, which desired nothing but seclusion, the Japanese family was even more patriarchal and disciplinarian than at the time of Pearl Harbour. It is obviously very useful propaganda to have psychologists prove that the enemy is mad. But as far as the truth is concerned, there can be no doubt that the idea that all readiness to fight is a semi-neurotic disposition, is completely wrong. The personal feelings of sedate psychologists leading comfortable lives are no guide in this matter. For a vigorous man, war may appear very attractive as an alternative to exhausting, monotonous work and grinding poverty. The 'heroic' narrative poetry from the *Iliad* and the *Niebelungenlied* to the *Mahabharatta* is full of glowing pictures of the life of warriors, amusing themselves with gambling, wine, women and song, and basking in glory, which stands in strong contrast to the abject fate of toilers. True, mechanization and discipline made war much less alluring. Nevertheless, it still has its attractions for many men. It all depends on what kind of life can be led in time of peace. If that is really abject, then most people prefer war: anything that promises a change.[1]

[1] The current notion that people fight better if they are well off economically is plainly contradicted by the facts. The whole of history indicates that those who lead miserable lives usually fight better than those who live in comfort, the explanation being that they value life less. This notion is used as an argument for proving that the Stalinist regime made its subjects opulent. If we accept this argument we should also maintain that the Japanese under Tojo were better off than the Americans or the French. Indeed, they ought to have been the richest people in the modern world, although in the past the Mongols of Gengis Khan ought to have been even richer. Ancient writers, like Polybios or Shang Yang, knew very well that rude rustics made

Another absurd idea is the notion that all racial and ethnic animosity is pathological: that it is some form of neurosis. Elaborate studies of anti-semitic or anti-negro personality are made, which purport to prove that it is the result of mental abnormality of some kind. I do not deny that sometimes it may be so; but as explanations of mass movements these studies are patently inadequate.[1] The command 'love thy neighbour as thyself', which was never considered by theologians as anything else but a counsel of perfection, is erected here into a criterion of mental abnormality; as if hate was not just as normal as love. Even if we accept the view that hate is the result of frustration, the fact still remains that there can be no life without frustrations: the essence of social life is not only mutual satisfaction but mutual frustration too. True, the pent-up aggressiveness is often unloaded on a scapegoat who had nothing to do with producing it. Usually, people choose for the scapegoat a foreigner in their midst. But again, the hostility of people having different customs and beliefs is surely natural; perhaps not in the sense of being innate, but in the sense of being the prerequisite of all culture. No culture is possible without normative codes, and these cannot be upheld unless deviations from them are condemned; therefore, foreigners who do not observe them must be looked down upon. The liberal intellectuals who think they are perfectly tolerant are mistaken. They are intolerant of religious intolerance, racial discrimination, tutelage of women, chauvinism

the best soldiers, and refined, comfort-loving city-dwellers—the worst. Actually, this idea underlies the whole sociological system of Ibn Khaldun.

[1] Even if we considered individual idiosyncrasies alone, it is anything but obvious that strong animosities towards certain groups are always due to neurotic dispositions. It all depends on the environment. Where such animosities form a part of the generally accepted ethos it is often precisely the neurotic who rebels against them. Thus, for instance, I found that among the several hundred South African students whom I taught, the few who were militant 'liberals' (in South Africa the term 'liberal' designates anybody who does not adhere to the policy of 'keeping the native in his place') showed unmistakably neurotic traits. Some of them were even obliged to undergo a psychiatric treatment. It might be argued, of course, that they became neurotic because they were persecuted by their fellows. But from the point of view of modern psychiatry this does not sound plausible, particularly as they showed little interest in the welfare work for the natives, which was done mainly by 'mild liberals'. On the other hand, if we meet a man in Sweden who declares that he hates all Siamese, and we find out that he has never seen one, then we have good grounds for believing that we are listening to a neurotic. There are, of course, many intermediate variants, but we must always remember that most neuroses are connected with the aversion towards the social environment, and therefore, we may expect to find a higher proportion of neurotics among those who reject the prevalent ethics than among those which accept it.

and many other things. They are indifferent to differences in language and race because they are adherents of the religion of Liberalism and Humanitarianism and participants in a supranational culture; and strong attachment to an ideology and the solidarity of believers can override the ethnocentrism. But perfect tolerance implies complete lack of moral principles. A certain amount of disdain for foreigners is, therefore, a correlative of ethnic distinctiveness. This consideration, however, explains only why the dividing lines in a conflict run in a certain way; it does not explain the occurrence of fighting because such hostility could be perfectly satisfied by a certain amount of isolation.

Everybody likes to feel superior to somebody. It is quite natural, therefore, that ethnic and racial distinctions should be fastened upon in order to satisfy this craving. Significantly, the stress on racial superiority is usually strongest among those members of the dominant race who have no other title to superiority. It is, therefore, quite natural that any sort of racial or ethnic or religious discrimination should be upheld by masses of people who, apart from economic benefits, derive great satisfaction therefrom. In doing this they are by no means necessarily irrational.[1]

But the causes of most violent frictions are economic. When there is not enough to go round there is a tendency to seize on any group distinguishing mark in order to exclude the competition. Persecutions of minorities were very often perfectly rational

---

[1] The word irrational has been greatly abused. Some people use it simply as exclamation or as a substitute for 'immoral' or 'ethically wrong'. Therefore, all attitudes of group antagonism are dubbed 'irrational'. This is incorrect. If I feel that Mr. X. is inferior to me simply because he is not me, and that he should be killed and his possessions given to me, then my feeling may be highly improper from the point of view of the Christian or Buddhist or Humanitarian ethics, but it is neither rational nor irrational. The same could be said about a feeling, which I may have, that all those who have short noses should serve those who have long noses. The terms 'rational' and 'irrational' are not applicable to feelings but only to actions and beliefs. An action is rational if the means adopted serve the purpose, and irrational if they do not. A belief is irrational if it is groundless, and rational if there are reasons for espousing it. Much muddleheadedness could be avoided if 'belief' were not so often confused with 'approval'. Expressions like 'do you believe in socialism?' are terribly misleading.

The term 'prejudice' has been also distorted. Correctly, it ought to be used to designate only judgment formed without consideration of evidence. If somebody believes that all Chileans are crooks, then that is prejudice; but if he dislikes them because of some characteristic which they really possess (say, because they are Roman Catholics), then this is antagonism—not a prejudice. True, antagonism and prejudice usually go together and reinforce one another; nevertheless, they ought not to be confused in a sociological analysis.

actions, due much more to economic conflicts, sometimes to plain greed, than to any psychological complexes.[1] And the intensity of these conflicts largely explains, as I shall show elsewhere, fluctuations in inter-group tensions.

To illustrate the point I can mention the example of South Africa. It is common knowledge that there is a definite, inverse association between 'the social background' and manifestations of the anti-Bantu attitude. The white artisans and labourers are much more violently in favour of 'keeping the kaffir down' than are the capitalists. The reason for this, however, is not any innate humanitarianism of the capitalists but the fact that their incomes and status are not directly threatened by the potential competition of the Bantu.

It is the desire to monopolize wealth, honours and other ophelimities, and to assure their transmission to the descendants, that calls forth the convictions of racial or ethnic superiority and inferiority. For this reason propaganda, adducing evidence that the potential ability of various racial groups is equal, can have only little effect on the acerbity of conflicts. These conflicts are not due to an intellectual error, but the possible intellectual error is the consequence of conflicting interests. This fact is so often overlooked because social scientists are still haunted by the concept of necessary universal harmony, elaborated by classical economists, according to which there can be no real conflicts of interests: all conflicts are due to misconception of interests. This, needless to insist, is an altogether chimerical notion.

In the light of the foregoing argument it is evident that some struggle must always go on in human societies; and if it is a struggle for the necessities of life it must involve killing. But why, we are bound to ask, do men always fight for the necessities of life? Why can they not just share them and live quietly? The answer to this has been given by Malthus.

The theory of Malthus has the privilege of being one of the very few sociological generalizations which possess the degree of certainty equal to that of the laws of physics. Indeed, its truth is no less certain than that of the statement that the earth is round. Its essence is simple in the extreme. Human population is biologically

---

[1] This statement does not, of course, imply that people begin to like or despise other groups only after deliberate calculation.

capable of doubling itself every generation, that is to say, about every twenty-five years. This, says Malthus, cannot happen because there is a limit to the amount of food any given territory, or the earth as a whole, can produce. Something must happen, therefore, either to the birth-rate or to the death-rate. Either some biologically realizable births are prevented, or people must live a shorter time than they are biologically capable of living. The factors which can lower the birth-rate—the preventive checks, as he calls them—resolve themselves into two categories: vice and moral restraint. He was mistaken as to the first because though prostitution and venereal disease may produce sterility, promiscuity as such does not. He was also too hopeful about the possible adoption of voluntary abstention, and did not envisage the modern practice of artificial birth control. Nevertheless, the central idea is irrefutable: either births are prevented or deaths must be more frequent than is biologically unavoidable. The factors which bring about the latter result—the positive checks—are three: war, epidemics and starvation; and ultimately they are all consequences of misery, that is to say, of the scarcity of food. And the only way of abolishing these is the prevention of births. In other words, high birth-rate must, in the long run, produce high death-rate, because population cannot grow indefinitely.

It is amazing that anybody endowed with a minimum of intelligence should attempt to disprove such an incontrovertible proposition. A simple calculation shows that even if we started with a single couple the biological powers of procreation would be sufficient to cover the whole surface of the earth with human bodies in a few millennia. Somebody has calculated that even at the present rate of growth, which is certainly below the biologically possible maximum, the population of the world would become so large before the lapse of two thousand years that there would be no room for people to stand on. Even if people inhabited many-storied houses covering the whole surface of the globe, even floating on the oceans, and lived on pills produced by direct transformation of solar energy, even then the end would be in sight because the increase in the mass of our globe would cause it to crash into the sun.[1]

---

[1] The views of Malthus have been and are being misrepresented in the most absurd manner even by people who are quite intelligent and even learned. Thus, for instance, Armand Cuvilier, the author of an on the whole excellent *Manuel de Sociologie* (Paris 1950), considers the fact that the population of Western Europe did not grow accord-

Until the introduction of modern contraceptive practices the importance of preventive checks was slight. All of them (abortion, abstention from intercourse, postponement of marriage, coitus interruptus) were so unpleasant that they would be practised only in case of dire necessity: when a substantial part of the population would be on the minimum subsistence level, and the positive checks in full action. In any case, the effects of their prevalence in some restricted circles would be outweighed by numerous births elsewhere. Killing one another could not have remained one of the chief occupations of men if there was no surplus of men available. And it was the natural tendency of the population to grow beyond the means of subsistence that assured the permanence of bloody struggles.[1]

The recognition of this fact enables us to advance a hypothesis about the origin of war. As nothing of that sort exists among the mammals this institution must be the creation of culture. It probably came into existence when the advance in material culture enabled man to defend himself better against the beasts which preyed on him, and thus to disturb the natural balance which keeps the numbers of any species stationary in the long run. After the beasts had been subdued, another man became the chief obstacle in the search for food; and mutual killing began. I was surprised to find that a similar view has been expressed by a Chinese philosopher Han Fei-tzu (*circa* fifth century B.C.) (quoted by J. J. L. Duyvendak in the introduction to his translation of *The Book of Lord Shang*,

ing to geometric progression to be a proof that Malthus was wrong. And that, when the whole *Essay* is devoted to showing that this cannot happen, except in an empty country; plainly, the writer in question never even glanced through it. It seems that Malthus is so often and so stupidly misrepresented because his theory touches the sex taboo, presents unpalatable truth useless for demagogic purposes, and in addition reveals the inevitable consequences of demographic expansionism, so alluring to group megalomania, and so dear to leaders lusting for power.

[1] Although sheer hunger drove men into battle much more often than people brought up in opulent countries imagine, the growth of population can produce war or some other form of strife long before the point of starvation is reached: the mere drop from the customary standard of living may generate bellicose pulsions. Moreover, intensive warfare may keep the standard of living well above the subsistence level. Malthus, commenting on the relative opulence of the Kirghiz, remarks: 'He who determines to be rich or die cannot long live poor' (*Essay*, p. 76). Many primitive tribes began to experience permanent starvation only after the pacification by the colonial governments.

I shall not dwell any longer on this point because the whole problem of the role of demographic factors in the causation of strife has been brilliantly analysed by Gaston Bouthoul in his books cited in the bibliography.

London 1928, p. 104) according to whom: 'The men of old did not till the field, but the fruits of plants and trees were sufficient for food. Nor did the women weave, for the furs of birds and animals were enough for clothing. Without working there was enough to live, there were few people and plenty of supplies, and therefore the people did not quarrel. So neither large rewards nor heavy punishments were used, but the people governed themselves. But nowadays people do not consider a family of five children as large, and, each child having again five children, before the death of the grandfather, there may be twenty-five grandchildren. The result is that there are many people and few supplies, that one has to work hard for a meagre return. So the people fall to quarrelling and though rewards may be doubled and punishments heaped up, one does not get away from disorder.'

The propensity of sex is inferior in motive power only to the need for food. And sometimes the means of satisfying it may constitute the object of the struggle. Female slaves were always one of the most alluring kinds of booty. Wealth and power, moreover, may be desired because they enable one to indulge in sexual pleasures. Their role in this respect varies according to the institutional framework. In societies which are steeply stratified and where women are chattels, rich and powerful men possess many women while the poor are often compelled to practise homosexuality or vent their lusts on animals; in Arab countries goats are widely used for this purpose. In such a situation wealth and power are more coveted than they would be if women could not be appropriated. True, even in a monogamous society wealth may be desired for similar reasons. Many a man in the Occidental countries dreams about becoming a 'big boss' and being able to afford smart and expensive mistresses. Also, a certain minimum income may be a pre-requisite of marriage. Nevertheless, there can be no 'cornering' of women in monogamous societies. Some additional pleasures may be inaccessible to the poor but no man needs to be deprived of a mate. A minimum income may be a condition of marriage in a monogamous society, but this is an absolute level which may be attained by the great majority, while in a polygynous society it is a matter of relative wealth: even if everybody were well-off, there would still be some richer than others, who would be able to 'corner' women. In such a society the struggle for wealth and power must be very acute.

In a situation where the richer the man the more numerous is his progeny, the unavoidable consequence is that some of them must be eliminated from the charmed circle, unless external conquests are taking place. Constant strife is the natural outcome of such circumstances. Sometimes the extermination of candidates to high office may be institutionalized: in the Ottoman Empire, for instance, a sultan was obliged by law to kill off all his brothers upon his ascension to the throne. In the Central African kingdoms of Ankole and Kitara the sons of the kings had to fight for the succession until only one of them was left alive. Sometimes, however, this process of elimination is peacefully regulated. In Siam the rank of the descendants of the kings (his successors excepted) was lowered after every generation; after the fifth they became commoners.

Extensive polygyny intensifies external and internal conflicts in other ways too. Firstly it accelerates the replacement of the killed, owing to the fact that under this pattern the birth-rate depends only on the number of women available. The result is that killings, i.e. wars and revolutions, must be more frequent. Secondly, polygyny, which is always practised more extensively by the upper strata than by the lower, and which therefore produces upward interstratic movement of women, accelerates the numerical increase of the upper strata at the expense of the lower; assuming that the ranks are inherited patrilineally, as they normally are in polygynous societies, because polygyny and patriliny are closely associated. If the upper strata are growing faster than the lower, then, in order to maintain their customary standard of living they must be continually raising their share of commodities produced by the latter, thus exacerbating the antagonisms between strata, or they must subjugate outside populations. The same situation may arise, though in a rather milder form, in a monogamous society in consequence of the lower mortality among the upper strata, due to their better conditions of living.

Christianity enforced monogamy as the best way of limiting sexual pleasures (of men, of course; women did not count), which were thought intrinsically evil. Its effects on politics were not envisaged. Nevertheless monogamy was one of the chief factors which made European political structures remarkably stable in comparison with Asia, and which made the internal politics of European states relatively free from violence. This was undoubtedly one of the chief causes of the distinctiveness of the Occidental civilization.

II

Stratification

How to estimate social inequalities

BEFORE proceeding to investigate the influence of military organization on social stratification, it is necessary, in view of the unsettled state of sociological terminology, to say a few words about the meaning of the terms used. A social stratum may be defined as an aggregate of individuals having more or less the same status in a given society. By status is meant the prestige enjoyed not because of individual peculiarities but in virtue of position occupied, i.e. the social role performed. I use the term 'aggregate' instead of the more obvious word 'group' because the latter implies that the relations between individuals constituting it are fairly close. This need not be so; as, for instance, in the case of medieval barons or Polynesian chiefs in sparsely populated areas. The only relation between its members which is indispensable for the existence of a social stratum is the mutual recognition of equal status. The consciousness of solidarity or antagonism towards other strata may or may not be present.

In spite of the simplicity of the definition, and of the fact that we all seem to know what stratum is, there is no dependable yardstick by which we could measure status. The forms of address and other visible forms of deference are obviously our sole indicators. But how can we tell when these are genuine and when spurious? An all-powerful dictator may be surrounded by a semi-divine halo, but those approaching him will call him 'comrade' in spite of their trembling knees. Or one may have to endure a waiter's impertinence in a British restaurant while being called 'Sir'. And 'Your obedient servant' of the official document may not perhaps be approached without gestures of propitiation. Did women have higher status fifty years ago than they have today, as one would expect judging from the readiness of men to bow and show other signs of deference?

Moreover, while in some societies, as in Russia in the eighteenth century, a definite system of ranks may exist, in others there may be no clear delimitation between the strata, as in the contemporary West. The presence or absence of such delimitations does not depend on inter-stratic mobility, which can be extremely intensive and general in societies and groups where ranks are rigidly demarcated, as for instance in the Ottoman Empire. For the benefit of those not acquainted with sociological terminology I must add that the term 'interstratic mobility' designates ascending and descending movements of individuals on the ladder of social stratification. Generally speaking, the greater the differences in status the stronger is the tendency for the corresponding patterns of behaviour to become institutionalized, which involves a clear demarcation of groups to which they apply. This is particularly likely to happen in societies not undergoing rapid changes.

I have been speaking so far about differences in prestige which various groups enjoy. The next step is to consider how are they connected with the distribution of wealth. If we classify groups or individuals according to their wealth we may find that this classification does not coincide with the one based on status. This problem was exemplified in a question over which one of my South African students pondered in his essay: 'is the status of a rich butcher higher or lower than that of a poverty-stricken professor?' Or one can think of the very common phenomenon of impoverished noblemen scorning the newly rich, who acknowledge themselves the superiority of blue blood. In spite of such exceptions, however, we can say that in the long run an impoverished group will also lose prestige, while that of the enriched group will be enhanced. But we should remember that we are dealing with an extremely abstract phenomenon; we may ask—prestige among whom? There may be conflicting claims which have not yet been settled by general acceptance. Every group tries to foster its material interests and if it does not succeed it means it has no influence, no power. Lack of wealth, then, proves powerlessness; and prestige is essentially determined by power. The discrepancies between prestige and wealth must, therefore, be considered as transitional phenomena, except in the case of groups whose claim to superior status is based on religious capacity requiring ascetism.

Since we are studying here the influence of military organization on social structure, it would be tautological to define political stratification as the grouping of individuals according to their political power. In order to be precise I must add that by political I mean the aspect of social organization which is concerned with the regulation of the use of violence. The term 'political stratification' will be used to designate the grouping of individuals according to their political rights. Political power may be exercised (i.e. the government may be influenced) in many ways: by passive resistance, unwillingness to fight, sabotage, bribery, moral reprimands, etc. Political rights differ from those channels of political influence in that they are claims to exert influence on the actions of the government, recognized by laws or customs. Generally speaking, the distribution of actual power tends to be reflected in the distribution of nominal rights, which, however, is also shaped by political ideas—perhaps a legacy of the past.

For the purpose of the present investigation I shall use the term 'social stratification' to describe the combined result of the evaluation of status, distribution of wealth and of political rights. In other words, inequalities in respect of wealth, status and political rights will be lumped together for the sake of the economy of words, and called social inequalities.

Contrary to the common preconception, social inequalities may be extremely sharp in a society where interstratic mobility is very great. Thus, for instance, in the Ottoman Empire at its apogee all the highest officials began their careers as slaves; the Delhi Sultanate in India even had a slave dynasty; Mamluks, the military group ruling medieval Egypt, were on principle recruited from foreign slaves. The Maori, on the other hand, provide an example of a society where, though everybody's 'station' was determined by birth, social inequalities were not very great.

Basic Causes of Stratification and Basic Forms of Power

There is a tendency nowadays, due to the influence of Marx, to think of the causes of social stratification as being mainly of economic nature. It is true that differences in wealth can arise even in societies devoid of political hierarchy, provided that there is something that can be appropriated. Among the Kazaks, for example, whose political structure is very rudimentary, the primary basis of distinctions is wealth in herds. Those who lose their cattle may be reduced to servitude. Even more striking is the case of the Goajiro Indians, living on the Carribean coast of Colombia. This tribe abandoned hunting, and went over to pastoralism after receiving cattle from Europeans. This was followed by the rise of inequalities based on differences in the number of cattle possessed. The neighbouring tribes, which did not adopt cattle breeding, show no traces of stratification. Here, then, we have the clearest possible case of a change in economy producing social inequalities. It should be noted, however, that these inequalities are rather small. I have not been able to discover a tribe where great differences in wealth would not be accompanied by a political machine of coercion, enabling the wealthy minority to protect their wealth, except where it is already protected by magico-religious beliefs.

Any large group needs co-ordinating organs. Very often co-ordination cannot be done through consultation, but only through subordination. In a large community, therefore, some hierarchic organization is unavoidable; particularly when it comes to fighting. Peoples who could not evolve or adopt such organization were inevitably destroyed. Now, certain privileges must be given to those placed in positions of command. A plentiful supply of able aspirants can be secured only if such posts carry some privileges of at least honorary nature. Honorary distinctions are in a way psychological correlatives of attitudes of obedience and command, and are inseparable from any hierarchic organization. Also, wealth always inspires awe. A hierarchic organization, the higher rungs of which were worse paid than their underlings, could not work. The common bent of the poorer to court the favours of the richer would undermine discipline. Moreover, in all groups performing tasks not in themselves enjoyable supervisors must be privileged. Otherwise, they would make common cause with those they are supposed to force to work. The following account well illustrates the point. The Polish army, which was reconstituted in Britain in 1940 with men who managed to get there by various ways after the occupation of Poland and France, contained far more officers then required. As according to Polish military law a rank could be taken away only as a punishment, the superfluous officers were incorporated into all-officer units where they performed the duties of ordinary soldiers. It was interesting to watch how the conduct of individuals changed when they were transferred from such units to formations where they had functions corresponding to their rank, or vice versa. When they were in positions of responsibility and honours they were far more severe with themselves, more eager to do their duties. '*La noblesse oblige.*'

In every large society, then, there must be some sort of stratification, and, other things being equal, the larger the society the higher the stratification. But it must not be forgotten that the size of a society is only one of the factors determining its stratification. The American society must be more highly stratified than a band of Semang, numbering a few dozens of individuals; but social inequalities in the kingdom of Dahomey are more pronounced than in America. The Buinese society, pervaded by differences of status and wealth, was not larger than the egalitarian League of Iroquois. The size of the society sets only upper and lower limits to the height of stratification.

In nearly all societies social inequalities go far beyond what is necessary for the smooth working of administration. Once the habits of obedience become established, it is easy for those in positions of command to use their authority to extend their privileges. How far they will be able to extend them depends on the balance of power between the rulers and the ruled. We can imagine a sort of a tug-of-war between the two groups. But we should beware of taking this schema for reality, where there is no clear cut frontier between the rulers and the ruled. Our imaginary tug-of-war symbolizes the result of countless human actions and attitudes; attempts to increase one's share of wealth or enforce more accentuated forms of deference; resistance to such attempts on the part of others, alliances and counter-alliances etc; all effected through countless forms of pressure and counter-pressure which human beings can exert on one another, ranging from physical violence to gentle disapproval. The result of these interactions depends on many circumstances: the number of contending groups, their relative power and cohesion, their constellation, the technique of supervision and possibilities of sabotage, ease of maintaining secrecy, etc. Nor should we forget the influence of ideologies. To a certain extent the beliefs about what are the proper privileges of any group are shaped by actual practice.

Generally, we are inclined to consider as just—we 'expect'—what normally happens. This is what Ihering calls 'the normative tendency of the actual'. Nevertheless, as I have tried to show in the article contained in the December 1949 issue of the *American Sociological Review*, the ideas must be considered as forces possessing considerable measure of independence.

The possession of wealth gives power. But as soon as we enquire into the meaning of the word 'possession', we see that the economic power is derivative. The terms: possession, property, ownership, designate the right to control, to use and dispose of objects, the access to which is prohibited to all except the owner. The norms, legal or customary, which support this control constitute, therefore, the foundations of economic power.

The basic rules of honesty are common to all peoples and form a part of the ethics of neighbourliness, indispensable to any kind of social life. Nevertheless, in highly stratified societies, consisting of a multitude of groups, frequently with divergent ethical norms, the poor often consider encroachments on the property rights of the rich not only as permissible, but even as meritorious. This popular opinion has found expression in tales, to be found all over the world, about good robbers who despoiled the rich and distributed their booty to the poor. Generally speaking, in societies where inequalities of wealth are extreme, the property of the rich is respected mainly because of the fear of punishment. Through early conditioning of generations, it is true, 'habits of honesty' can be inculcated, which are observed almost automatically. Nevertheless as the experience of revolutions shows, once the notion spreads that one can help oneself to the goods of one's richer neighbours with impunity, these habits disappear quickly.

We see then that economic power is not self-sufficient but derivative. On the other hand, the ability to compel through the use or the threat of violence is an irreducible form of power, which can exist without being supported by anything else. Spoliation is the usual fate of wealthy groups which do not wield political power. Plutocrats of Greek cities, Italian and German bankers in the epoch of the Renaissance, Chinese and Japanese merchants throughout history, provide examples to the point. The situation in the countries of Islam is well brought out in the following passage taken from the work of the greatest social philosopher of the

Middle Ages—Ibn Khaldun (*Prolegomenes Historiques*, t. 2, Paris 1936, p. 293): 'A burgess who possesses enough money and goods to be regarded as the richest man of the town draws upon himself envious looks, and the more he displays his wealth the more he exposes himself to be harassed by emirs and princes. . . . And when by some chicanery they manage to convict him of some offence . . . they take away his riches. . . . Everybody, then, who possesses great wealth . . . should have a protector who can shelter him . . . a member of the royal family, or one of the favourites of the ruler . . . or a leader of a party, strong enough to make the sultan respect him.' The proverb 'who pays the piper calls the tune' does not hold if the piper is stronger and can rob the payer, as is shown by the outcome of the dealings of various financial magnates with Hitler. It is true that sometimes businessmen exercised considerable influence, even though they did not control the government, because they were indispensable. Thus the medieval lord obtained a share of the wealth produced by his burghers but he could produce none himself; he had, therefore, to give them various liberties.

In view of these facts it is not surprising that it is almost always those who wield the military power who form the supreme stratum of society. The pure plutocracy, that is to say, the rule of the rich who do not control the military power, can only be a temporary phenomenon. Purely economic factors produce, no doubt, fluctuations in the height of stratification, but, as the following evidence will show, the long-term trends are determined by the shifts of the locus of military power.

In all social conflicts violence is the argument of the last resort. Even where it is never used it stands in the background as the enforcement of the 'rules of the game'. In a strike or a lock-out, business competition or electioneering, no violence may be used, but it is so because the police threaten with violence anybody who would use violence against other contestants. There are, of course, limits to what compulsion can do: it can enforce the grudging acquiescence and the execution of assigned tasks, if this can be effectively supervised; it cannot produce willingness to put out maximum effort or readiness to make sacrifices, and it stifles initiative. These consequences of a social order based on compulsion may have a very unfavourable influence on the quantity and the quality of production, thus leading to general impoverishment, which, however, need not affect the ruling group if it can alter the distribution of wealth even more in its favour. The passive resistance of the masses is most dangerous to the ruling group when the state is fighting for survival. At such times, if the willing co-operation of the masses is militarily essential, an effort must be made to win them over, to convince them that they are fighting for themselves. And the rulers who have to convince the masses to this effect may end by convincing themselves that their task is to serve and defend the People. For these reasons, the technical and military circumstances, which make the willing co-operation of the masses in the war effort more or less essential, are the most powerful among the factors which determine the extent of social inequalities.

Economic power is derivative, but the same cannot be said about power based on magico-religious beliefs. As a matter of fact, this power seems to be the earliest foundation of social inequalities. Many extremely simple societies, such as, for example, various Siberian tribes, with no traces of chieftainship, have shamans enjoying high status, unusual wealth and other privileges. Among the Trobriand islanders the power of the chief has a magical basis. He possesses, it is true, considerable wealth, acquired through polygynous matrimonial arrangements, which enables him to display generosity indispensable for maintaining his influence. The ultimate weapon, however, against disobedience and encroachment on his prerogatives is his sorcery, which is believed by the commoners to be of deadly efficiency. The power of the Church in medieval Europe was based on the firmly rooted belief that it was the intermediary between God and mankind. The wealth, which the Church accumulated in consequence of its power, far from being an element of strength, was a root of weakness. Its accumulation led to laxity among the clergy which undermined the people's devotion, and then the riches of the Church became an easy booty for princes and nobles. Perhaps the most striking example of the efficacy of magico-religious beliefs as a foundation of power is the supremacy which the Brahmins maintained in India for two thousand years. It is particularly remarkable that they, unlike the Egyptian or Catholic priesthoods, were never organized.

These few examples are sufficient to show that the magico-religious power, unlike the economic, is irreducible to any other form. For this reason we must make allowances for this factor, when studying the influence of military organization on social structure, particularly where the magico-religious influences are, as in India and Polynesia, overwhelming.

The regime of castes in Bali is perhaps the most startling example of how religion can enforce a gradation of deference independent of, and cutting across, the differences of wealth and authority.

War and Social Inequalities

Herbert Spencer formulated a generalization that militancy, i.e. the orientation of a society towards war, conduces to greater social inequalities. In support of this contention he gives numerous examples of rigid distinctions of ranks in societies which he considers as militant, like Russia or Germany, while pointing out to the lack of such clear-cut distinctions in the industrial society of Britain. We need not dwell on the utterly inadmissible identification of 'industriality' with peacefulness. But it must be said that it is by no means obvious that the definiteness with which ranks have been delimited varies, as Spencer assumes, with the span of inequalities between them. Although the distinction between nobles and commoners is quite clear cut among the Buinese, such inequalities as that between a millionaire and a pauper in the fluid Western Society are completely unknown to them. In the history of Rome we find that the obliteration of legal barriers between plebeians and patricians was followed by a widening of the span of inequalities. Two classes hitherto non-existent came into being: the extremely wealthy and the destitute. The 'clients' had originally a status, rights and duties defined by law. During the later period of the Republic all legal distinctions between them and their patrons disappeared, but their position deteriorated instead of improving: they became a rabble of flatterers and hangers-on.

Herbert Spencer's proposition, therefore, must be split into two questions which must be answered separately. First, does militancy as such necessarily produce a widening of social inequalities? Second, is it conducive to their definiteness?

The empirical evidence shows that protracted wars sometimes widen social inequalities, but sometimes have the contrary effect.

28

The wars of the late Roman Republic are an example of the former, but those of the Early Republic—of the latter. The great intensification of warfare witnessed in the twentieth-century Europe produced a definite flattening of the social pyramid. Many societies constantly engaged in ferocious warfare never possessed any permanent stratification, e.g. various tribes of American Indians.

Success in war, more than in any other human activity, depends on co-ordination of individual actions, and the larger a group the more necessary is the co-ordination, and the larger the hierarchy required. We should expect, therefore, that the larger the group the more pronounced should be the stratifying effect of militancy. Even the most intensive warfare between very small groups can hardly produce a stratification, except if it leads to conquest. Conquest has a doubly stratificatory influence: in the first place it directly creates a division between the conquerors and the conquered; secondly, it opens new possibilities of inequalities by bringing into existence large political units which can be organized only on hierarchic lines.

We must remember, however, that social inequalities may have other than military foundations, so that an intensification of warfare may not heighten the stratification but merely give more importance to military as compared with civilian hierarchies. That is what happened, for instance, in U.S.A. as the result of the two World Wars.

Moreover, the intensification of warfare may make it necessary to enlist the support of the masses by granting them various privileges, in which case a substantial levelling may take place. The necessity of such a course will depend mainly on whether mass armies are, in view of the state of tactics and armament, more or less efficient than professional armies.

Security from invasion may enable a military aristocracy to maintain its position even though the state of military technique permits the use of mass armies. The Japanese, for instance, were acquainted with the Chinese military technique, employing large conscript armies. But they did not adopt it because they had no large frontiers to defend. Bearing of arms remained in Japan an aristocratic privilege, with the exception of the short period of Taikwa reforms. The result was that the supremacy of the military aristocracy was never challenged.

Hierarchic organization implies, as we saw, differential privileges. These privileges, however, cannot exceed certain bounds without impairing the fighting spirit of the army. The ordinary warriors must feel that they have something to gain by victory, or at least something to lose by defeat. Never has it been found profitable to use starved, ill-treated drudges as soldiers. Moreover, the example of great conquerors shows that a substantial measure of equality in the distribution of booty and the enjoyment of comforts enhances the morale of armies.

The morale of the troops gains in importance as the struggle becomes more ferocious and its stakes higher. It is significant that never in the history of European armies was the chasm between officers and men deeper than in the eighteenth century; and that was precisely the time of limited wars, mildly conducted for insignificant stakes. As the wars became more ferocious the armies became more egalitarian. Nowhere can it be more clearly seen than in the case of Britain after the end of its 'splendid isolation' in 1914: before that date the gentlemen officers were divided by an impassable barrier from other ranks, recruited mainly from the paupers. The army which fought the second World War was completely different: the differences in pay and status were not extreme and ascension from the ranks into the officer corps was quite usual.

A fact, well known to all those who participated in any war, must also be mentioned: namely, that there is far more equality between various ranks on the battlefield than in the barracks. This can be anecdotally illustrated by the sentence which came from the mouth of a certain professional officer: 'The war is over; now we can make an army out of this rabble.' It is true, of course, that on the battlefield there are fewer amenities which can be invidiously distributed; but also, a strong feeling of solidarity inevitably spreads among those engaged on a common, arduous, and above all dangerous task. This applies not only to military units but to whole nations too. The feeling of ardent solidarity—with the concomitant projection of all hatreds on the enemy, makes the holders of wealth and privileges more willing than they would be otherwise to share these boons with their less fortunate countrymen. For the same reason nationalism, even in its aspects not directly connected with war, may, by generating a feeling of solidarity in all members of the nation, bring about the same result. Increase in militancy, therefore, may, contrary to what Spencer thought, have a levelling, and not a stratificatory, effect.

A clear-cut demarcation of ranks has been found necessary not only in armies, but in all large organized groups: it is a correlative of their hierarchic structure. The indeterminateness of social stations, like that prevailing in modern Western countries or in Rome during the Late Republic and the Early Empire, is due to the multiplicity of autonomous groups and to the existence of numerous individuals independent in many ways. Such societies may be termed poly-hierarchic. In mono-hierarchic societies, like, for instance, ancient Japan, where there is only one hierarchy embracing the whole society, the strata tend to be strictly delimited. To the extent then, to which intensive warfare may lead to the regimentation of individuals hitherto enjoying considerable freedom and to the absorption of all other hierarchies by the state bureaucracy, and of the latter by the army, we may say that, other things being equal, militancy does promote the definite demarcation of strata.[1]

Conquest and the Origin of Privileged Groups of Warriors

A privileged group of warriors who alone bear arms may arise in two ways: it may differentiate itself gradually from the rest of the population, or it may impose itself through conquest. The

[1] Social structures differ in the degree of what might be called 'delineatedness'. In some of them the divisions between various groups are clearly demarcated, so that there can be no doubt about where any given individual belongs; in others the dividing lines are blurred. This distinction applies, naturally, to stratification. The ranks of a modern army represent a delineated stratification, whilst the class divisions of any contemporary Occidental society provide an example of blurred stratification. It must be remembered that this 'blurredness' is an objective fact—not a consequence of the defectiveness of sociological knowledge. For this reason if we attempt to calculate exactly the class-position of an individual in a society with a blurred stratification, as some American sociologists do, we are trying to locate something that does not exist.

Within one collectivity some groupings may be delineated, others blurred: in the British House of Commons, for instance, the party groupings are delineated but the groupings into cliques are blurred.

This distinction should not be confused with that between intersecting and mutually exclusive groups. Thus, for instance, sporting clubs are usually clearly delineated groups but need not be mutually exclusive: one can belong to a tennis club and a rowing club at the same time. On the other side there are mutually exclusive groups which, nevertheless, are not clearly divided one from another, e.g. cliques into which members of many university senates form themselves.

These distinctions are very important for the understanding of the functioning of social structures, but their analysis would exceed the bounds of the present enquiry.

first process may occur either because the restriction of military service leads to the attainment of maximum strength, or because some groups try to monopolize arms-bearing in order to obtain a privileged position. That, however, may happen only in already fairly complex societies, where costly armament beyond the means of many may render the services of the majority useless, or where internal and external security are such that disarming the population becomes feasible. Where hunting and war weapons are substantially the same, and hunting remains economically important, such disarmament is impossible. That is why the earliest forms of privileged warrior group arose, no doubt, through conquest.

It would be tedious and unnecessary to enumerate even a minute fraction of the examples of this process which have been noted since the dawn of history. Most of them are rather too complicated for the present purpose. When we look, for instance, at the Roman conquest of Gaul we see a subjugation of a society, already highly stratified, by another whose stratification is even more complex. The best authenticated cases of the rise of privileged groups of warriors out of the collision of unstratified peoples can be found in Africa; particularly in East Africa and Sudan, where extensive kingdoms have been founded through the conquest of negroid agriculturists by hamitic pastoralists. Although some of these states are now ethnically homogeneous because of the early amalgamation of the two races (e.g. Ganda), in others the miscegenation was not very extensive, or at least was not accompanied by social coalescence. In the kingdom of Ankole, for instance, the barriers between the two groups were in full force until the imposition of colonial rule. The Bantu agriculturists were compelled to pay tribute to their masters and allowed no political rights. Significantly, they were debarred from participation in wars, which was an exclusive privilege and duty of the pastoral aristocracy. As weapons were extremely simple, it was impracticable to disarm the serfs completely; nevertheless, the lack of training was a sufficient guarantee of their impotence.

The foregoing example is peculiar because of its analytic purity: we can see there how the conquest produced stratification in the absence of any other possible causes. But conquest always has stratificatory influence. The conquered are always disarmed and the conquerors always reserve for themselves the right to bear arms. These propositions are almost self-evident and the examples supporting them range from ancient Assyria to present-day South Africa. Nevertheless, it must be remembered that these propositions apply only to complete conquest, and not to various shades of dependence, such as vassalage, protectorate, etc. Moreover, the conquerors may recruit soldiers from among the more reliable elements of the population, particularly if the country is heterogeneous and its unity is merely nominal, without being rooted in group consciousness. In India, for instance, the Sikh and Gurka troops constituted the pillars of the British rule. In spite of these qualifications, however, the fact remains that in states founded upon conquest, and where the ethnic amalgamation did not take place, the proportion of warriors in the total population is determined less by what is the optimum from the technical point of view, than by the numerical relation of the conquerors to the conquered.

Military Participation Ratio

We have seen that militancy sometimes heightens and sometimes flattens the pyramid of social stratification. Which way its influence will be exerted depends mainly on whether the co-operation of the masses is essential for the successful prosecution of war or not. It depends, in other words, on the proportion of militarily utilized individuals in the total population. This proportion I shall call the military participation ratio. The question now arises: what determines it? But before we answer the question we must make a distinction between the actual and the optimum military participation ratio. The actual M.P.R. is the one which actually obtains in a given society. It influences directly the stratification by affecting the balance of power. The optimum M.P.R., on the other hand, is the ratio which, within the given technico-tactical conditions, would enable a state to attain the maximum military strength, other things, like morale, leadership, etc., being equal. Obviously, the actual and the optimum M.P.R. do not always coincide, and the changes in social stratification during or after wars are often due to the fact that the war emergency compels states to approach the optimum M.P.R. and to abandon the previously existing actual M.P.R., if it deviated from the optimum.

But what determines the optimum? The first factor which must be taken into account is the relation of the cost of the most efficient armament to the productive capacity of the country. It is the relative, not the absolute, cost that matters. The equipment of a medieval knight would be considered nowadays very cheap in comparison with all the gadgets used by contemporary soldiers. But in the Middle Ages it was altogether impossible to provide any substantial portion of the inhabitants of any country with the equipment of a warrior. Even the possession of horse was beyond the means of an ordinary peasant. And as the heavily armed cavalry could disperse any number of footmen, the only alternative to defeat was the institution of a small stratum of professional warriors whom the rest of the population would support. It must be remembered, however, that at any time various sorts of armament can be produced, some more, others less costly; and that a great host poorly equipped may be stronger than a few superbly equipped warriors. Moreover, it is possible to combine better or worse, lightly or heavily, armed troops in varying proportions. Without going into the intricacies of this problem, it is sufficient for our present purpose to realize that the cost of armament together with the relative merits of various possible constitutions of armies circumscribe the optimum M.P.R.

If the skills necessary for a warrior can be acquired and kept only by protracted and continuous training, military service tends to become professionalized. Normal economic activities cannot be reconciled with continuous military training except in the case of primitive hunters. The intensification of agriculture, connected with the growth of the population, requires steady application. Also if the wars are waged in distant regions—distant, that is to say, in relation to the development of the means of communication —the professionalization of the army will be fostered.

Equipment and Remuneration

The influence of the changes in M.P.R. on stratification will be modified according to whether the warriors equip and maintain themselves or whether they are equipped, provisioned and paid by the government. Naturally, in the first case they will be much more independent, much better able to enforce their claims. Moreover, if they are equipped and maintained by the government they are, as a rule, also fitted into an army organized by the government and commanded by officers appointed by it, and not left in their territorial or kinship units under the command of their tribal or elected leaders. They can thus be disciplined and their aspirations much more easily disregarded. The armed forces of the ancient Greek and Latin city-states, of the Swiss cantons, and of nearly all primitive tribes belong to the first category; the armies of bureaucratic states, like ancient oriental monarchies or the states of modern Europe, belong to the second. High M.P.R. will exert a stronger levelling influence if the armed forces fall into the first category. I shall come to some further implications of this point later.

The Facility of Suppression

The influence which the M.P.R. exerts on the span of social inequalities is further modified by another factor which we must also examine.

The degree of superiority of the armed forces over the unarmed populace depends in the first place on the quality of the armament: a machine-gun confers upon its possessor a much greater superiority over an unarmed crowd than a sword. Also, if the production of armaments can be undertaken only in large-scale establishments, it is impossible to prepare an uprising secretly. Swords or even rifles can be manufactured clandestinely but not tanks or bombers. We can say, therefore, that the predominance of the armed forces over the populace grows as the armament becomes more elaborate. This predominance can also be due, sometimes entirely, to the advantage of organization. The importance of organization varies. In medieval Europe, when wars were waged by disorderly hosts casually assembled, it was slight as compared with Graeco-Roman and modern times. The preponderance of medieval knights was based on the exclusive possession of costly arms and not on organization. The importance of organization and the difficulty of improvising it grows as the numbers involved increase. We should expect therefore, that, other things being equal, the larger the state the more difficult is rebellion. That is perhaps why Montesquieu thought that small states tend to become republics while large ones are usually monarchies. Generally, we may say that the helplessness of the populace against armed forces grows with the importance of organization.

Privileges of a ruling group are limited by, among other factors, the police technique with the aid of which it can keep the masses down. These techniques are not, as many erroneously think, something quite new. It is true that never before has there existed a police apparatus equal in scope and efficiency to the Gestapo or M.V.D. (the latest name of the Soviet police). But most of their basic methods are very old and seem new to the inhabitants of western Europe only because they were not used there for a few generations. The Venetian and Spartan oligarchies had in their service innumerable spies, intricately organized. Chandragupta, the founder of the first empire in India, used a whole army of spies and 'agents provocateurs'. Quite advanced methods of torture were known already at the time of Hammurrabi and it does not seem that much progress has been made in this field since then. Terror was employed by Ivan the Terrible and Hai-ling Wang with skill hardly surpassed by the contemporary masters of the craft. Aristotle gives a list of means used to maintain tyrannies, in which figure: the principle of divide and rule, extensive espionage, fostering licentiousness so that members of a family would denounce each other, terror and deceit. The superiority of contemporary police systems rests on new technical means with which these old methods can be applied, such as the radio, the telephone, the card-index system, finger-prints, etc. These new means make it possible to spread the web of espionage and terror over greater areas and envelop in it bigger populations. As far as the intensity of terror is concerned it seems that the Moscow of Ivan the Terrible reached the summit of what is humanly possible. Nevertheless, the extension of the police systems brought about their strengthening. In olden days many successful revolts started in the outlying regions which were beyond the reach of effective supervision by the despot. In a modern totalitarian state there are no such sheltered spots. The possibilities of rebellion are, therefore, correspondingly reduced.

The ease with which a population can be kept down, which depends on the circumstances just discussed, is a very important determinant of social structure. As I shall need to refer to it frequently later, I propose to call it 'the suppression facility'.

The suppression facility tends to sharpen social inequalities. It accentuates the effects of the low M.P.R. and counteracts the effects of the high.

Frictional Factors

The factors which determine the optimum M.P.R. undoubtedly exert very strong influence on social structure, but they do so only in so far as they modify the actual M.P.R.; and the actual M.P.R. does not adjust itself automatically to the optimum, as it may be enmeshed in various extraneous circumstances which, following the usage of the economists, we may call frictional factors.

In a comparatively isolated society, such as Egypt before the Hyksos invasion or Japan before Perry, the need to attain the maximum strength may not be urgent: the military organization may, therefore, deviate considerably from the optimum possible. Also, the existing stratification may offer resistances to the changes required for bringing the military strength to the maximum. The lower classes may be so oppressed and full of hatred against the government, that the latter may shrink from arming them whatever the circumstances. Besides, because of apathy, resulting from age-long oppression, such a step might be useless; the peasants of the Late Roman Empire often preferred Teutonic invaders to imperial tax-collectors. In states dominated by privileged groups of warriors, particularly in those based on conquest, the resistance to the extention of military service may be especially strong, because such extension might mean also the extension of privileges, and therefore the diminution of benefits. Such a reaction is exemplified by the resistance of the Japanese samurai to the creation of a westernized army based on the conscription of peasants.

Conversely, an egalitarian society where all men bear arms may resist the introduction of more efficient methods which make universal military service useless. The Old Testament depicts such resistance offered by the Israelites to the military innovations of David, Solomon and other kings, who found that heavily armed charioteers were necessary in the struggle against their neighbours.

Sometimes the social structure of some country may be particularly suited to new forms of military organization, which can be developed there earlier than elsewhere, thus giving that country predominance out of proportion to its numerical and economic strength. This was the case of Sweden in the seventeenth century, which was the first state in Europe to have a national army because it had never known fully developed feudalism.

Technico-military prerequisites of national mass armies existed in eighteenth-century Europe, but as this type of army was incompatible with the aristocratic structure of the European nations,

37

it was not adopted. Armies were based on recruitment of mercenaries and on limited long-term conscription. These armies were rather inefficient instruments of war, but the wars were waged for limited objectives, and therefore there was no need to adopt an uncongenial form of military organization. The possibility of raising mass armies was first exploited by the French revolutionary government because this type of army harmonized with the new society. Other states did not adopt this new form of military organization until faced with extinction.

Owing to the fact that the defeated are usually disarmed, every conquest tends to lower the actual M.P.R.; and as, generally, states try to subjugate as much territory as they can, there is the constant tendency for the actual M.P.R. to be reduced below the optimum. This, incidentally, is the explanation why tremendous empires often fall prey to small tribes where all men bear arms. Moreover, any group which is victorious in the struggle for power, which goes on in every society, is inclined to buttress its privileges by reserving for itself the monopoly of arms, thus reducing the actual M.P.R. below the optimum. Nevertheless, when faced with the threat of extinction the states usually adopt the most efficient military organization, even though it may be uncongenial to their social structure. If they do not then they are swallowed by those who have done so. In this way the technical and tactical innovations, which alter the optimum M.P.R., affect the actual M.P.R., and modify social structure.

The Influence of the Standard of Living

The ultimate results of levelling tendencies depend on a basic economic factor: the wealth per head, which is determined by the relation between the size of the population and the productive capacity of the country. When there is not enough food to go around, a struggle literally for life between individuals and between groups must ensue. Those who get to the top cannot afford to be generous, particularly if they are prolific, as that would mean starvation for them or their children. Moreover, if there is a strong pressure of population on the means of subsistence, an increase in the food supply would merely enable a great number to survive without improving their lot. Punishments for crimes must be severe as only the threat of torture or death can stop a starving man from stealing or robbing. No political system based on compromise can survive in such circumstances, as can be seen from the fate of the attempts to introduce parliamentary forms of government in the poor countries of eastern Europe, Latin America and Asia. Tensions generated by over-population and the consequent poverty can be relieved by foreign conquests and migration. Even then, however, social structure would be affected by becoming attuned to war. The better struggle for existence and opulence, not merely for power, will not permit the narrowing of the chasm between the ruling and the ruled. The dispersion of power in consequence of the extension of military service cannot in such circumstances result in a permanent measure of equalization of political rights and wealth, but only in periodic outbursts of levelling rebellions. This is what happened in China, as will be shown later, while in Europe of the nineteenth century the growing wealth enabled the shift in the balance of power to find its expression in the more stable form of the democratic state.

B. EMPIRICAL VERIFICATION

Primitive Peoples

The so-called primitive peoples exhibit an astonishing variety of structures, ranging from small bands to kingdoms comprising up to a million inhabitants. In stateless tribes, i.e. those which do not possess differentiated organs of government, all men are warriors. Sometimes, it is true, only the young men are permanently at arms on the forward posts while the married men are in the reserve, as among the Masai of East Africa. The prestige accorded to warriors depends, generally speaking, on militancy. Among the Indians of the North American Plains, where prestige was gained almost exclusively in virtue of combative prowess, ageing led usually to its loss. Typically gerontocratic tribes, like the Australian, are not as a rule very warlike. In spite of the great variety of military arrangements, in no simple tribe is there a differentiated stratum of warriors. Such a stratum comes into existence only in larger societies—states which, without exception, are founded upon conquest. As examples I can mention Nupe in Nigeria, Ruanda in the Congo and Ankole in East Africa. But even conquest does not necessarily lead to the formation of a stratum of warriors. The Zulus of South Africa, for instance, incorporated all the defeated tribesmen who made a submission, into their own tribe, which thus grew to a tremendous size. The same can be said about various political units which rose and fell among the Eurasian nomads.

The arms of primitive tribes are usually too simple to force the M.P.R. down. There are exceptions to this rule. Thus, for instance, when bronze swords percolated to prehistoric Europe, their costliness strengthened the position of the chiefs, who alone could afford them. European rifles and cars played the same role in Arabia and North Africa. Nevertheless we can say that restriction of military service in primitive societies is due, as a rule, to the effect of conquest and not to the cost of armament.

It is unnecessary to dwell long on the stratificatory influence of conquest: the facts are plain. There are no records of a case where the conquerors acquired no privileges. Superimposition of one group over another is a facet of conquest. However, it is very seldom that we know the past of a pre-literate people, and for this reason it is mostly impossible to trace the changes which primitive societies underwent when some of them conquered others. Even in cases where it is reasonably certain that a conquest did take place, legends are usually our sole source of information about the conditions preceding the conquest. Therefore, as far as preliterate societies are concerned, we are compelled to abandon nearly all attempts to study variations in time, and resort to static, intersocietal comparisons.

On principle, we could try to ascertain a coefficient of association between steep stratification and low M.P.R. I have no doubt that such an investigation would be valuable, but I do not feel able to undertake it unaided. The labour involved in tabulating would be immense and, I am afraid, out of proportion to the results which could be obtained through it; because no matter how carefully we counted the instances, the delimitation of the units would remain largely arbitrary, thus rendering the exactitude spurious. Furthermore, the counting would depend on the availability of information; and we have no grounds for believing that the peoples about whom we do have information, constitute a representative, random sample; indeed we have a definite reason to hold a contrary opinion, owing to the unevenness of the geographical distribution of our data. So, we are condemned to operate with rough approximations in all matters of quantification.

In order to forestall misunderstandings I must repeat that I do

not maintain that the M.P.R. is the sole determinant of stratification. Such a contention would be absurd and could easily be
disproved: there are many societies in Polynesia, Indonesia and
Melanesia where all men are warriors, and where the stratification, being rooted in religion, is very steep and rigid. The restriction of bearing of arms is only one of the roots from which
social inequalities can spring. But, on the other hand, the evidence
of primitive peoples undoubtedly attests the proposition that the
restriction of bearing of arms (i.e. low M.P.R.) widens social inequalities. In some ways, this evidence is even more conclusive
than that of complex societies, owing to the absence of complicating factors operating among the latter. In fact, in spite of a protracted search through ethnographic literature, I have not discovered a single pre-literate society where the existence of a group
of warriors, possessing the exclusive right to bear arms, is not
accompanied by a rigid division into the ruling warriors, enjoying
various privileges, and the ruled non-warriors, labouring for them.
As there are no negative instances, it would be almost a waste of
time to enumerate the positive. Whether we take the Mbaye in
the Chaco, or the Ruanda in the Congo, or the Fulbe states in the
Sudan, everywhere we find that those who bear arms constitute
a privileged stratum.[1]

In virtue of the fact that the armament of primitive peoples is
as a rule so simple that the optimum M.P.R. is always high, the
actual M.P.R. is lowered only in order to keep down those who are
disarmed; there can be no need for lowering it in the interests of
military efficiency. Consequently, the steepness of stratification is
an even more necessary accompaniment of a low actual M.P.R.
among primitive than among complex societies.

The Near East

The earliest theocratic city-states of Mesopotamia do not seem
to have had a differentiated stratum of warriors. Their armies,
simply armed and fighting in a sort of a primitive phalanx, had all
the features of a militia. These societies were not egalitarian,
although some traces of primitive tribal democracy survived for a
very long time; but the stratification was based on the religious
beliefs, bestowing great authority upon the priests, not on the

---

[1] I must add that these states have been built through conquests, and that the
stratification there corresponded to ethnic divisions.

military organization. Privileged strata of warriors arose only after the conquest of one city by another, and later of the whole Sumer by foreigners. It might be that the expensiveness of the bronze armament, which at that time became indispensable, also contributed to elevate the status of warriors by restricting their number. The changes in military technique and in the methods of production of armament which took place later did not affect the actual M.P.R.s of various political formations which rose and fell in these lands, because they were all founded upon conquest and preserved the character of conquest states. That is why the cheapening of armament, caused by the advent of iron, did not have a levelling effect, as it had in Greece and Rome. The Assyrians, who introduced iron armament into this area, super-imposed themselves as warrior-masters upon other peoples and disarmed them. Henceforth until World War I the overwhelming majority of the inhabitants of this area lived under the heel of foreign conquerors.

The fate of Egypt was different. The unified kingdom was built through conquest but no traces of ethnic distinctions were left in the historical period. The Hyksos, who invaded Egypt in the middle of the second millenium, were the first foreign conquerors of Egypt. They did not remain its masters for long but their rule left profound traces; the kingdom of the Pharaohs became a military state. Until then it had felt itself immune from attack. Its military organization was little developed. There were royal guards, partly Nubian slaves and mercenaries; the monarchs (provincial governors) possessed similar formations. There were military colonies on the border. But in principle all the inhabitants were liable to military service, though the levies were never general but only partial. They did not form a very efficient army by Asiatic standards but until the Hyksos invasion no emergency occurred with which casually assembled troops could not cope. Armament was of the simplest—spears, clubs, swords, small shields and small ineffective bows; tactics were practically non-existent. The Hyksos showed the Egyptians the invincibility of the Asiatic tactics, the main feature of which was the employment of chariots. The peasants could not, of course, be made into charioteers. The country could afford only a very restricted number of them. Moreover, all soldiers had now to be clad in protective armour, and protracted training became necessary. In such circumstances armies recruited through extensive conscription became useless, and were replaced by professional soldiery; for the first time in the history of Egypt a privileged stratum of warriors enters upon the scene. The chasm which gradually grew between them and the rest of the population became accentuated by the large scale recruitment of foreigners, because the Egyptians in virtue of their age-long pacification made bad soldiers. The status of the peasants was even further depressed: in addition to the priests and the court they had to work now for the soldiers, whose profession became hereditary.

Egypt remained under the dominance of warrior strata until quite recent times; but later it was the foreign conquerors and not indigenous soldiery who were the masters. The only successful peasant revolt in the whole history of Egypt was the great revolution which in the middle of the third millenium brought down the Old Kingdom and destroyed the aristocracy. Significantly, it occurred when the peasants were still eligible for military service, which they did not become again until the reforms of Mehmet Ali at the beginning of the last century. These reforms effected a substantial levelling of stratification by abolishing the privileges of the ruling warrior group of Mamluks.

Ancient Greece

The Greek city originated as a conqueror's castle. War-bands which in successive waves swept down the peninsula established themselves in fortified camps, usually on hill-tops, from which they dominated the subject populations. In Homeric times,[1] however, many of the cities were already ethnically homogeneous. Nevertheless, everywhere the warrior aristocracy remained sharply divided from the peasants. The reason why the ethnic amalgamation was not followed by the attenuation of social inequalities is not far to seek: 'The princes', says V. G. Childe (*What Happened in History*, London 1942, pp. 151 and 163), 'owed their power and wealth to a monopoly of new implements of war—long rapiers of costly bronze, huge shields and light horse-drawn war-chariots. . . . Battles resolve themselves into single combats between richly-armed champions. . . . These decide the issue; the infantry are

[1] Though the Trojan War was waged in the twelfth century B.C., we can be sure that the social conditions described in the *Iliad* were those of Homer's own time, i.e. the eighth century.

mere spectators. In fact only the few could afford the long blades of bronze, the chariots—marvels of the wainwright's skill—and the highly trained steeds, so that the masses were militarily worthless and accordingly politically impotent.'

'Cheap iron democratized agriculture and industry and warfare too,' says the same author. But we must add that the effects of it were not felt until certain tactical developments took place. In the meantime, social inequalities and antagonisms became even sharper.

The Dorian invasions brought into being a number of states, chief of them being Sparta, which did not follow the general lines of the social evolution of Greece. The M.P.R.s of these conquest states were determined not by technico-tactical factors but by the numerical ratio of the conquerors to the conquered. The proverbial Spartan discipline and training were the means of compensating for the paucity of numbers.

The breakdown of tribal organization, the spread of commerce and above all of money-lending, opened new possibilities of amassing wealth and of depriving others of it. The kings who often protected the plebeians against the nobles disappeared. Their loss of power seems to be due mainly to the fact that the monopoly of external commerce, which was their chief source of income, became impossible to maintain when the passive trade with the Phoenicians was replaced by sea-going expeditions of the Greeks themselves. The expansion of commerce and the growth of cities did not, it must be emphasized, in themselves produce democracy as can be seen from the following passage written by the famous French historian (Glotz, *La Cité Grèque*, Paris 1928, p. 103): 'Broadly speaking the new economy rapidly swelled the ranks of the lower classes and aggravated their condition. As the rich became richer, the poor became poorer. . . . Usury ground down small men . . . the plebeians of the town . . . lived from hand to mouth on wages which the increasing use of that human chattel, the slave, was forcing down.' The conflict between the strata was naturally violent. But what demands an explanation is: why did the scales turn against the oligarchy and in favour of the masses in the sixth century?

The answer is given by Fustel de Coulanges (*La Cité antique*, 28th ed., Paris 1923, p. 326), who says: 'During the first centuries of the history of the cities the strength of armies resided in the cavalry. The true warrior was the one who fought on a chariot or on a horse; a foot-soldier was of little use and therefore little esteemed. All ancient aristocracies reserved for themselves the right to fight on horseback . . . gradually the infantry became more important. The progress in the manufacture of arms and the birth of discipline enabled it to resist the cavalry . . . soon it began to play the first role in battles because it could manoeuvre more easily. And the legionaries and hoplites were plebeians . . .' Aristotle formulates a general theory in this respect (*Politics*, Everyman's ed., pp. 191–5): '. . . those who are employed in war may be likewise divided into four; the horsemen, the heavy-armed soldiers, the light-armed, and the sailors; where the nature of the country can admit a great number of horse, there a powerful oligarchy may easily be established: for the safety of the inhabitants depends upon the force of that sort; but those who can support the expense of horsemen must be persons of some considerable fortune. Where the troops are chiefly heavy-armed, there an oligarchy, inferior in power to the other may be established; for the heavy-armed are rather made of men of substance than the poor; but the light-armed and the sailors always contribute to support a democracy . . . so that for an oligarchy to form a body of troops from these is to form it against itself.' 'The first states in Greece which succeeded those where kingly power was established, were governed by the military. First of all the horse, for at that time the strength and excellence of the army depended on the horse, for as to heavy-armed infantry they were useless without proper discipline; but the art of tactics was not known to the ancients, for which reason their strength lay in their horse: but when the cities grew larger, and they depended more on their foot, greater numbers partook of the freedom of the city.' Significantly, even in Athens it was only when the fleet became the basis of its might that those too poor to afford the equipment of a hoplite, but whose services as oarsmen now became essential, gained equal rights. After the Persian wars, the Athenian state became a sailors' republic. Thessaly, on the other hand, where the cavalry continued to be the principal arm, was never affected by democratic movements.[1]

During and after the Peloponnesian war the situation became

[1] Significantly, in the Carthaginian republic, which, after the sixth century B.C., relied almost exclusively on mercenary troops, the oligarchy was never threatened by

complicated because Athens, Sparta and later Thebes constantly attempted to impose their own constitutions on other cities, so that these depended less on internal circumstances than on the balance of power.

The Macedonian conquest altered the situation completely. Mercenary armies were the pillars of royal authority in the Hellenistic kingdoms. Social inequalities even increased and the hatred of the poor for the rich certainly did not disappear. Nevertheless, there were no more confiscatory laws because the fortunes of the rich were now guarded by the kings' mercenary soldiers, often foreigners.[1] The royal treasury represented now the main danger to private wealth.

Iran

The social stratification of the Achaemenid empire was rather complex. There were serfs, the majority presumably descendants of the pre-Aryan population, and the nobility, mostly descendants of the Aryan conquerors. But there were also nomadic Aryan tribes possessing no serfs. Furthermore there was a gradation among tribes, some being considered nobler than others. Moreover, within the Aryan tribes themselves certain lineages acquired privileged positions. There was no absolute gulf between the nobles and the common freemen who were quite numerous, and lived in self-governing villages. Whether they were more or less numerous than the serfs cannot be determined. The army of the Achaemenid kings was also mixed. It contained feudal cavalry but also contingents of free commoners, horsemen and foot-soldiers, led by their tribal chiefs. The main arms were the bow and the long spear. Protective armour was scanty and uncommon; even the Royal guards had none. With the conquest by Alexander, the alien ruling group imposed itself upon the Iranian population.

The Sassanid dynasty, which rose to power in the third century A.D., rejected all the vestiges of the Hellenistic dominance and emphasized the Iranian heritage. The Aryans and the original

popular movements, though the economic situation was similar to that of commercial cities of Greece.

[1] This statement applies to the cities under the direct control of the monarchs, though even there abortive revolutions took place. In the independent or semi-independent cities revolutions continued to occur; for Sparta the third century was the age of revolution.

inhabitants having long before coalesced, there were no ethnic divisions of great importance. Nevertheless, social inequalities became steeper than in the earlier periods. Apart from the few merchants and artisans residing in towns, society consisted now of the nobles and the serfs, and between them there was an absolute, impassable barrier. The disappearance of the common freemen was no doubt connected with the changes which took place in the military sphere. With the introduction of the stirrup heavy protective armour came into use. Owing to its expensiveness it was far beyond the reach of ordinary commoners—even of those among them who could keep horses. The mainstay of the army was now the heavily armed cavalry. Only noblemen, feudal vassals of the king and his bodyguard, could serve as horsemen, and as the infantry composed of impressed serfs was purely auxiliary, they were the only section of the population which had military power. No wonder then that they were able to acquire various privileges and reduce the peasants to harsh servitude.

After the fall of the Sassanid dynasty Iran remained under the rule of foreign conquerors until the fifteenth century. Arabs, Seldjukid Turks, Mongols, Timurid Turks succeeded each other as masters of Iran. Even during the reign of Safevid dynasty (sixteenth to eighteenth centuries), power seems to have lain in the hands of various Turkish and Afghan tribes. The Iranian peasants and herdsmen remained militarily helpless and therefore downtrodden.

China

Not much is known about the structure of the Shang kingdom. Possibly it resembled somewhat the theocratic city-states of early Sumer. But however that may be it is certain that after the Chou conquest the country was dominated by the military aristocracy. The feudal society which came into being was in many ways similar to medieval Europe. Nevertheless, the predominance of the nobles over the common population never reached the same pitch. The Chinese noblemen of the Chou era resided in walled 'cities' in considerable groups headed by the princes, and not in individual manors like their European counterparts. The common people lived in village communities. They had to pay to their lords dues in kind and in labour but they were free from the constant interference which the European serfs had to suffer. Moreover, and this is truly remarkable, in this feudal society an ideology spread which stressed the duties of the lord towards his subjects. This ideology reflected, it seems, the state of the balance of power between the nobles and the commoners, which was, when compared with the situation in medieval Europe, much more in favour of the commoners.

The picked troops consisted of armoured chariots, manned by the nobles, from which they fought with bows and spears. But every chariot was surrounded by a division of foot-soldiers fighting with javelins, short swords and bows. 'The most common arm was the powerful reflex bow, which was able to pierce the best armour worn by aristocrats. This meant that the aristocracy was not in the position of the knights of Europe, who could don their armour, bestride their chargers, and go out against any number of armed peasants or foot-soldiers and laugh at them. Even dukes and kings of ancient China were frequently wounded by arrows. These facts lead to two conclusions. First, whenever the populace became really disaffected, it could and did simply desert the ruler in the face of the enemy, and he was powerless. Second, if the people were oppressed to the point where resentment was universal, they possessed the power to revolt, and the aristocrats could not stand against their united opposition. Unlike the war lords of modern China they had no machine-guns with which a handful of mercenaries could mow down thousands of peasants. For this reason then it was essential that the people should be kept contented or at least below the line of active resistance against their superiors' (H. G. Creel, *The Birth of China*, London 1936, p. 364).

During the period of 'warring kingdoms' iron weapons became common and the feudal levies were replaced by mass conscript armies. Leading such an army Shih Huang-Ti conquered all other principalities and founded a unified empire. This change, of course, was not followed by the advent of democracy. Nevertheless social inequalities perceptibly diminished. The aristocracy lost their privileges. Their serfs became owners of the land owing dues to the state only. Members of the ruling bureaucracy were recruited among the lower strata.

Except under the dynasties of conquest, when bearing arms was reserved for the conquerors, the Chinese armed forces remained a mixture of conscripted peasants which were called up in emergencies only, of criminals impressed for long-term service, of mercenary shock troops (often foreigners), and of military colonists similar to the Roman *limitanei* or the Russian cossacks.

The fact that the armed forces were based in so large a measure on conscription made the distribution of power very different from what it was in the Islamic Near East or in India. It was this fact which made the so-called dynastic cycle possible. Chinese history comprises several such cycles. There essential stages may be summarized as follows: the degeneration of the dynasty enables the officials to get out of hand; they oppress the people and compel the free peasants by means of usury and chicanery to become their tenants; popular discontent coupled with the extreme incompetence of the government lead to a revolution; the new dynasty founded by a revolutionary leader proceeds then to abolish debts and redistribute lands, thus making the peasants once again the owners of the soil they till, and to replace the corrupt and incompetent officials by new officials, more honest and vigorous, largely drawn from the lower strata; after some time degeneration and corruption set in once again and the cycle begins anew. This is combined with the demographic cycle: the increase of the population during the prosperous periods aggravates the tension between the strata; the loss of life during the revolution raises the standard of living in the early days of each dynasty, as soon as the material damage is made good, which in view of the extreme simplicity of the productive equipment cannot take long.

Since the time of the unification until the T'ang dynasty the cavalry was steadily gaining in importance because the invasion of the nomadic riders proved beyond any doubt its superiority. It was also becoming more heavily armed because of the improvements in the technique of manufacturing armour and the adoption of the stirrup invented by the Eurasian nomads around the beginning of our era. The Chinese economy did not allow the peasants to equip themselves with horses and other paraphernalia of a horseman. Moreover, fighting on horseback is a very difficult art to acquire, far more difficult than driving a tank or even piloting an aeroplane; and it demands constant practice. The only solution was to have recourse to mercenaries, recruited mainly among the nomads. The peasant masses lost in military importance and, as we should expect, their position worsened. The dynastic cycle has not disappeared, but under the T'ang dynasty, which more than any previous one relied on foreign mercenaries, the span of social inequalities seems to have been wider than ever before. After the T'ang the situation seems to have remained without much change of any importance from the present point of view; except under the dynasties of conquest which henceforth ruled China most of the time, and under which the mass of the Chinese were disarmed and kept in an inferior position.

The contemporary events have so far deviated from the traditional cycle less than is commonly supposed.

Japan

Japan's past, though full of peasant rebellions, differs radically from that of China; these rebellions never succeeded. To be sure there was quite a number of successful revolutions but these were carried out by the nobles. The explanation of this fact must be sought in the form of military organization. Here the basic fact is that the bearing of arms remained throughout the Japanese history the prerogative of the nobles. The only exception was the time of Taikwa reforms in the seventh century, when the emperors, attempting to build a bureaucratic monarchy on the Chinese pattern, also tried to imitate the Chinese military organization. This part of these reforms proved just as unsuccessful and short-lived as the rest. Japan, not being threatened by anybody, did not need large armies and soon the nobles re-established their monopoly, which they maintained until the westernization of Japan. The result was that their supremacy was never challenged. There can be no doubt that the structure of the Japanese society would have been radically different if the external danger had made the use of large conscript armies unavoidable. The profound social transformation which occurred after Perry's expedition was due, perhaps, mainly, to the fact that it became necessary because of the Western threat to Japan's independence. Naturally, Japan did not become a parliamentary democracy, nor did it lose its highly stratified character. Nevertheless, contemporaneously with the adoption of semi-universal conscription many of the disabilities suffered so far by the lower strata, such as serfdom, lack of freedom of movement, discriminating taxation, distinct dress, etc., were abolished. Here again then we see that the rise in M.P.R. coincided with considerable levelling of stratification.

India

India is a classic land of subjugations; with the exception of the Near East no other land has been conquered so many times by foreign invaders. The rise of the 'caste system' [1] is principally, though not entirely, due to this fact. Another consequence of it was that armies remained strictly professional except among independent tribes and sectarian polities; the peasant masses remained permanently disarmed and, as we might expect, invariably downtrodden; and, in contrast to their Chinese counterparts, they have never succeeded in carrying out a large-scale revolution.

Little is known about the organization of the societies which belonged to the civilization of the Indus Valley. In the Indo-Aryan societies the profession of arms was reserved for the Ayran conquerors. With the rise of the royal power the character of the armed forces changed, except in the republics which survived till the late Mauryan era, where the warrior nobility ruled the serfs. These republics resembled in many ways the rural 'city-states' of ancient Greece, such as Thebes. Elsewhere royal soldiers replaced the knights.

In all the kingdoms about which we have information, even in those under strictly indigenous dynasties, the armies were professional. The Mauryas and the Guptas possessed very large armies, comprising detachments of infantry, elephants, chariotry and cavalry. The soldiers were either paid from the imperial treasury or assigned lands on semi-feudal tenure, their position being somewhat similar to the soldiers of Hammurabi. Although the 'caste system' has never been so rigid as it appears in the Brahminic apologetics, and neither all warriors were Ksatryas nor all Ksatryas warriors, in most of the kingdoms the soldiers did tend to form a hereditary stratum.

The situation in the kingdoms of the Pallavas, Cholas, Vijaynagar and other southern principalities resembled in all essential

---

[1] An enormous amount of paper has been wasted on worthless theories about the 'caste system', and on speculations about whether there are 'castes' outside India, by writers who did not realize that there is no such thing as 'caste' in India. 'Caste' is a word coined by the Portuguese who did not understand the Indian order very well, and this word is indiscriminately applied to both 'varna' and 'jat', which are totally different things, neither of which corresponds to 'caste' as defined by textbooks of sociology. The strata of the Iranian society under the Sassanids provide the best example of the latter.

aspects the conditions in the Mauryan and Gupta empires. Only the kingdom of the Rashtrakutas deviated from the pattern: there the soldiers were drawn from all strata, and did not form a hereditary stratum. However, nothing resembling universal military service was in force: the army was definitely professional. Nevertheless, this widening of the field of recruitment considerably enhanced the strength of the kingdom, and, as we should expect on the basis of our theory, was connected with the general weakening of the caste barriers, which revealed itself particularly in the relatively small number of the pariahs and their status, less depressed than elsewhere.

Indigenous kingdoms continuously attempted to swallow one another, and from time to time extensive empires were erected in this way. But apart from this there was a constant intrusion of conquering peoples from the north. In the states which came into existence in consequence of such conquests, the profession of arms was, naturally, reserved for the conquerors, and the subjugated populations were disarmed. In the kingdom of Harsha the Sakas —a tribal league containing peoples related to the Huns of Attila —constituted a ruling stratum of warriors. In the Sultanate of Delhi the Arabs and the Afghans, in the Mughal Empire the Turks, Persians and other adventurers occupied equivalent positions. There were also smaller Radjput states, which flourished especially when the great empires were in the state of collapse, where warrior nobility ruled the serfs.

The only places in India where social inequalities were smaller were the inaccessible regions occupied by independent tribes. Some of these tribes, particularly those in central and southern parts of the country and in Assam, were, and still are, very primitive and lacking in social differentiation beyond that based on sex and age. But the mountain tribes of northern areas are generally large, possessing hereditary chiefs and nobility. Nevertheless, the commoners are not downtrodden. On the contrary, they are renowned for their pride and independence. The chiefs are more leaders than absolute rulers on the southern pattern. The explanation of this deviation from the pattern of stratification prevalent further south is to be sought in the military organization. In the mountain tribes every man is a warrior. He has his own rifle and is imbued with a sense of honour.

The case of Sikhs is equally illuminating. They were a people of soldiers, and refused to recognize the barriers of either 'varna' or 'jat'.

The picture, described by the Greek travellers, of the Indian peasants tilling their fields while the armies were fighting is by no means a reflection of a rustic idyl; on the contrary, it is an indication of utter political apathy of the peasants, induced by the complete disregard of their welfare on the part of the ruling strata. To the peasants all rulers were equally bad.

It is quite possible that these military circumstances, by producing an all-pervading mood of hopelessness, had something to do with the fact that India and the Near East were the cradles of messianic, other worldly religions. These religions in turn became important forces, moulding societies.

Rome

Before the Etruscan conquest the inhabitants of Latium lived under a tribal organization similar in many ways to that of some African tribes. They were ruled by the elders of the 'gentes' (which correspond to what ethnologists call clans) and an elected chief. There already existed relations of dependence, servants and retainers came to be attached to individual joint families, but the sharp differentiation into nobles and serfs arose only after the Etruscan conquest. The Etruscans together with some Roman families formed the aristocracy. 'The masses of the native population'—says Roztovzev (*Social and Economic History of the Roman Empire*, Oxford, 1926, pp. 11–13)—'were forced to toil and sweat for their new masters. The overthrow of the Etruscan dynasty by the aristocracy of Rome did not alter the prevailing economic conditions. Much more important was the need of maintaining and developing a strong military organization able to defend her from attacks coming from the North and from the rivalry of other Latin cities. . . . It was during this darkest period in the history of Rome that the foundations of the Roman peasant state were laid. How and when the former serfs of the aristocracy became free peasants, owners of small plots of land and members of the plebeian class, we do not know. It is probable that there was . . . a gradual evolution bringing with it both an emancipation of the former serfs and an increase in the number of free peasant landowners, who had never disappeared from Roman economic life, even in the times of the Etruscan domination. Both developments are probably to be explained by the military needs of the Roman community. . . . The Servian reform . . . was both the consecration and the formulation of the results of an economic and social process which took place in the dark fifth century.' But why should the military needs be different in the fifth century than before? The explanation is the same as in the case of Greece: The development of tactics and discipline made numerically large armies useful. The legendary 'celeres' of Romulus were nobles on horseback. Iron became sufficiently common by the fifth century for the fabrication of fairly cheap arms. The 'classis', the Roman counterpart, though by no means a replica, of the Greek 'phalanx', made the foot-soldiers supreme on the battlefield. The enrolment of all citizens able to equip themselves became the condition of survival.

The economic factors cannot account for this levelling. On the contrary, as in Greece, they opened new possibilities of exploitation. The development of commerce did permit the rise of the new group of non-patrician rich; but it certainly did not strengthen the position of the peasants who constantly pleaded for laws which would prevent their exploitation: above all the mitigation of the law of debt. As warriors they found a weapon in the refusal to fight. When the patricians invoked their aid, their leaders said: 'They have stripped us of our farms, they have carried our brethren into bondage, and now they call upon us to fight their battles.' The aims of the plebeians differed according to their economic position. To the rich the most important were the right of admission to magistracies and the right to associate with the patricians and intermarry with them. The poor desired mainly the abolition of debts, the cessation of the enslavement of debtors, a share in the conquered lands and the codification of laws, which would protect them from their arbitrary interpretation. Both desired the right to participate in making laws. These rights were won during the period from the fifth to the third century. It is true, of course, that there was never equality of political rights; the voting was weighted heavily in favour of the rich.

Significantly those who were too poor to equip themselves for military service were practically deprived of the vote. And, what is equally important from the point of view of this study, the rich in whose favour the vote was weighted were those who served in the cavalry and heavy infantry, requiring costlier equipment than the minor sections of the armed forces. The reorganization of the army effected in the fourth century and ascribed to Camillus abolished all differences in armament in the heavy infantry of the legion, and this was probably another circumstance fostering social levelling.

'The great annexations of territory of the fourth and the beginning of the third centuries,' says Leon Homo (*Roman Political Institutions*, London 1929, pp. 54–5, 59, 71, 83)—'following on the subjection of Latium and the unification of peninsular Italy, led to a considerable extension of the citizenship. . . . The gradual conquest of Italy increased the power of the army inside the Roman state, and therefore that of the plebeian element, which was the backbone of the army. . . . In the interests of the country plebs, who chiefly wanted material betterment, the nobility in power took very effective measures. In 328 the Lex Poetelia Papiria . . . ordered the abolition of seizure of the person . . . and the liberation of men imprisoned for debt. Many colonies were founded, chiefly for the benefit of the rural plebs. . . . The great constitutional reforms of the third century, representing the triumph of the peasant middle classes, launched the Roman state in the direction of a regime of rural democracy. . . . In the first part of the third century it was not the Senate—as formerly in the fourth century and later from the Second Punic War—which had supreme direction of foreign policy, but the people, won over to . . . ideas of expansion.'

The earlier part of the third century marked the crest of the democratic trend. Henceforth, its driving force, the free peasantry, began to disappear. The cheap Egyptian corn, exacted as tribute, undermined their economic position, already difficult in consequence of devastating wars. Victories brought tremendous numbers of slaves, thus making *latifundia*, based on gang-slavery, profitable. Many peasants became heavily indebted to newly enriched tax-farmers and officials, laden with the spoils of provincial rule. Political rights were also affected: the Senate became once again the closed and preponderant body. The army was at the same time becoming professional. Wars in distant regions could not be waged by conscript armies. Moreover, the tactics were becoming more involved and required therefore longer training. After the reforms of Marius the army ceased altogether to be a militia of Italian peasants and turned into professional, long-service troops, paid and equipped by the generals.

The four processes: the ruin of the peasants, the professionalization of the army, the increasing disparity of wealth and the concentration of political authority in the hands of oligarchy— went together. I do not maintain that the professionalization of the army was the cause and the other three effects; that would be too one-sided. But it is evident that without it the other three tendencies could not develop as they did. It was possible, after all, to prevent the ruin of the peasants by legislative means, as was done in the fourth century. The attempts of the Gracchi to prevent the ruin of the peasantry failed, because the peasants, being no more militarily indispensable, could not exert the same pressure as before.

The peasants were abandoned but the urban proletariat was pampered. 'Panem et circenses' was the demand which even the most despotic dared not refuse. But it was from among these proletarians that the army was recruited. They in turn began to be neglected in the later period of the empire when they ceased to supply recruits, which were now drawn from the outlying regions of the empire and even from outside. The peasants remained downtrodden.

Byzantium

Apart from the areas only partly subdued by the imperial government, the cultivable lands of the Byzantine Empire were occupied by two different kinds of rural organization: the great domains and the village communities. The former were tilled by the serfs and, occasionally, by slaves. The latter comprised legally free peasants, owning the land they cultivated, who were jointly responsible for their taxes. Both the great domains and the village communities survived throughout the life-span of the empire but their relative importance varied. Variations in this respect implied transmutations of serfs into free peasants and the reverse, and they were connected with the transformations of military organization.

During the four centuries which followed the foundation of Constantinople, most of the lands of the empire belonged to the owners of the latifundia. These magnatees, ruling sometimes thousands of serfs, were not feudal vassals of the Emperor but merely landowners; because, though they maintained considerable numbers of armed retainers, they did not receive their lands as the reward for military service. Their military importance was secondary. Mercenaries of variegated provenience formed the mainstay of the armed forces.

In the seventh century the onslaughts of the Persians, Arabs, Slavs and Eurasian nomads almost annihilated the empire. In order to build an army able to repel the assaults, with which the mercenaries (owing primarily to their paucity) could not cope, Emperor Heraclius and his successors reformed the military organization. They confiscated many large domains, freed numerous serfs, and distributed lands to those who were willing and able to fight. The recipients of the lands were obliged always to be ready for service in the army. The resulting system was a kind of peasant militia which, however, was led by regular officers of the government. This system enabled the Byzantine Empire to survive and even to reconquer a considerable part of the lost lands.

In the eleventh century this system began to decay. It was not, however, replaced by a disciplined professional army, as of old, but by armed forces which, though containing some regular soldiers, resembled the feudal levies of the medieval Occident. The great landowners and their retainers came to play the most important part in the defence of the empire. But they played this part very ineffectively, and the abeyance of the previous system, involving the reduction in the number of soldiers, contributed probably more than any other circumstance to the downfall of the empire. From the point of view of the present enquiry, the most interesting fact is that this transformation of the military organization, involving a lowering of M.P.R., was accompanied by the widening of social inequalities. The independent peasants almost disappeared, the great majority of them having been demoted to serfdom, and the area occupied by great domains underwent great proportional expansion.

North Africa

There are three types of geographical environment in North Africa: the mountains, the arid steppe and the cultivable plains; and each is, or at least was until the advent of the motor-car and the aeroplane, the cradle of a different type of society. Comparing these societies we notice again the close connection between stratification and M.P.R. The deserts are occupied by the tribes of pastoral nomads, led by the sheiks who are leaders rather than rulers. Within these tribes a rough equality prevails, tempered only by the privileges of seniority and personal prestige. All men are warriors. Sometimes one tribe established a supremacy over others, but this never amounted to a true domination.

The inaccessible plateaux and mountains were the seats of the minute republics of the Berbers. These republics possessed councils of elders who settled current matters. Nevertheless the final instance was a general assembly of heads of families—that is to say, men who were not under the authority of their fathers. The economic equality was almost complete, and all men were warriors.

On the cultivable plains the situation was completely different. There, the disarmed peasants worked for their masters: the sultans, their relatives, officials and mercenaries.

The frontier between these types of societies fluctuated, and there were many transitory stages. Nevertheless, the division was as clearly marked at the beginning of this century as it was in the days of Carthage.

Medieval Europe in General

The lands of the Roman Empire were ruled with the aid of mercenary troops. During the lengthy process known as the fall of the Empire, these armies were becoming more and more Germanic and less and less under the control of the imperial government. Finally, the Germanic warriors usurped the place of the Roman ruling class. Naturally they regarded the right to bear arms as their exclusive privilege. The evolution of the military technique continued to proceed in the same direction as during the later period of the empire: towards the increasing predominance of heavily armed horsemen. This was mainly due to the introduction of the stirrup, invented by the Eurasian nomads, which enabled a rider protected by heavy armour to fight efficiently. We cannot be sure in the case of western European lands that the military service has been restricted because the armament became costlier, as this restriction was a legacy of the Roman past, and moreover the Germanic invasions would tend, as usually in such cases, to narrow or at least maintain this restriction. Nevertheless, it can be said with certainty that the medieval military technique would prevent the extension of military service even after the ethnic fusion took place. It has been calculated that several years' income of the whole village was necessary to cover the cost of equipping one knight. And nothing could oppose a warrior clad in protective armour on an armoured charger. So, ineluctably, the peasant masses were militarily useless and helpless. This explains why all peasant revolts have, without exception, been drowned in blood, and the lot of the peasants became worse after every such attempt. Medieval armies, unlike the troops of post-Ch'in China or of modern Europe, never went over to such rebels because the nobles themselves were soldiers. With the significant exception of Scandinavia, peasants possessed no political rights, which were nevertheless accorded to townsmen who could defend themselves against the feudal cavalry behind their walls.

Medieval Germany

When the Germanic tribes first came in contact with the Romans, their stratification was very slight. All men met under arms and ordered the affairs of the village, the district and the tribe. They divided fields, passed resolutions about war or peace and chose the leader. Although certain families enjoyed special consideration, there was neither priesthood nor nobility with definite privileges. But henceforth social inequalities grew steadily. In the case of Germanic kingdoms established in previously Roman territory this was largely due to the effects of conquest. But even among the Germanic peoples which remained in their native lands social inequalities continued to grow. After the end of the Great Migrations, the common freemen lost their political rights. More and more of them were depressed to the status of serfs, until the society came to consist of serfs and nobles. These transformations, we can be sure, were not unconnected with the lowering of M.P.R., caused by the innovations in armament. The Germanic warrior of the earliest times was armed with a wooden club, a short spear with an iron point, eventually a battleaxe and a sling. Horsemen had neither saddle nor stirrups. Gradually and continuously the armament was becoming heavier, until after Charlemagne the heavily armed knights attained undisputed predominance on the battlefield, and the tribesmen in arms were replaced by feudal hosts.

Poland

The history of Poland exhibits a similar connection between the cost of armament and social stratification. The original kingdom was despotic; and, although the nobility already existed, its privileges and its power over the common people were not very extensive in comparison with the later times. The armed forces consisted of the general levy of freemen and the king's permanent guard—druzjina, corresponding to Carolingian antrustiones. Both were armed with very primitive weapons and used no protective armour. Their method of resisting the hosts of heavily armed German knights, whom they could not oppose in a pitched battle, was ambush—tactics well suited to a country still covered by primeval forest and morasses. Gradually, the general economic progress enabled Polish warriors to equip themselves better. The Polish forces which defeated the knights of the Teutonic order on the field of Grünewald were similar to the hosts of western Europe: their mainstay were the heavily armed horsemen. The social structure of the Polish kingdom underwent a corresponding modification: the nobles succeeded in interposing themselves between the common people and the crown. All peasants were reduced to the status of serfs, while their obligation to participate in wars disappeared. Their primitive weapons, the only weapons they could afford, were now useless. The heavily armed horsemen remained the mainstay of the Polish armed forces right down to the end of the seventeenth century; and, significantly, Poland was the country in which the privileges of the nobility were greater than anywhere else in Europe.

Sweden

Sweden provides us with an example of the opposite development. There, the densely wooded nature of the country and also the lack of suitable horses reduced the cavalry to unimportance. Until the reforms of Gustav Vasa the armed forces consisted of the general levy of freemen if the country was invaded, and of volunteers from all strata for foreign expeditions. Infantry was far more important than the cavalry. The history of the Swedish military organization thus deviated from the general European pattern and, significantly, its social history is also strikingly different from that of the rest of Europe: there was no feudalism in Sweden. The tremendous majority of its inhabitants remained peasant-proprietors, who owed allegiance to no lord. The main difference between the nobles and the peasants was that the former possessed more land. No judicial or administrative powers over the peasants were vested in the nobles, except in those who were at the same time royal officials. The peasants were judged by tribunals composed mainly of their peers and took orders only from the king. In short, there was no serfdom in Sweden. That country, moreover, was the only one in Europe where representatives of the peasants sat in the parliament.

Curiously, the cavalry began to play a greater role in Sweden precisely at the time when its preponderance was waning in western Europe, that is in the fourteenth century, and tax exemptions were granted to anybody who could equip himself with the outfit of a horseman. This trend, however, did not last sufficiently long to produce important modifications of stratification, although the tendency towards the widening of the privileges of the nobility revealed itself during this period. The trend was arrested when Gustav Vasa, building on old foundations, introduced conscription for foreign service and organized the first national army of Europe.

Denmark and Norway

Until the thirteenth century the social structure of the Scandinavian states remained essentially similar; they all differed from the other countries of Europe in more or less the same way. After the thirteenth century social inequalities became greater in Denmark than in Sweden, in which latter country they became greater than in Norway. This ranking corresponds to the relative importance of the peasant militia in these countries. In Norway it remained the mainstay of the armed forces, owing to the nature of the land, even less favourable to the employment of cavalry than in Sweden; in Sweden the cavalry became more important, though, as we saw, it never became nearly so important as it was in the rest of Europe with the exception of Switzerland and the Scottish Highlands; the armed forces of the Danish kingdom resembled those of the rest of Europe, and, as we should expect on the basis of our theory, social inequalities became there almost as great as they were in Poland: towards the end of the Middle Ages Denmark was fast becoming a seignorial quasi-republic.

Switzerland

Switzerland was the only country in medieval Europe which was never ruled by the nobility. The free peasants were never subjugated: there were no serfs. Throughout the Middle Ages and later, their communes governed themselves according to the rules of the ancient Germanic tribal democracy. Even the war leaders were elective and vested with their authority only for the duration of a campaign. The geographical situation was the cause of this survival of tribal democracy amidst countries dominated by feudal lords. In the mountains the lightly armed Swiss foot-soldiers were invincible. Moreover, they evolved the infantry tactics which, when they spread to other countries, precipitated the demise of the feudal cavalry. On this invincibility were based both the country's independence and the military organization in the form of militia, which in turn was reflected in the political institutions and the social structure.

Russia

The eastern Slavic tribes have been welded into a state by the Norse conquerors, who became the ruling stratum. They lived in fortified places, some of which grew into commercial towns, and drew tribute from the Slavic peasants. Although they engaged in trade, principally by exchanging the proceeds from the tribute against Byzantine manufacturers, they remained essentially warriors—the prince's 'drujina'. The ethnic amalgamation and the disappearance of the trade, the latter being due to the incursions of the nomads, led to a change in the character of the ruling stratum: they ceased to live in towns, supported by the gifts from the prince and proceeds from the trade, and began to dwell on lands assigned to them by the prince in return for military service. The peasants were liable to various kinds of corvee, some of which were of military nature, but the cavalry, consisting of nobles, formed by far the most important part of the army. And so it remained until fire-arms became really important. In the meantime, though the M.P.R. did not undergo any perceptible changes, the position of the nobles has been transformed: they fell into much stricter subordination to the tzars. This was the exact opposite of the process which was going on in Poland at the same time.

The fire-arms brought into existence a new kind of warriors called 'stryeltzy'. They were foot-soldiers recruited from outside the nobility, and rewarded with beneficia. They were the essential instruments of the tzars in their struggle against the nobles. Gradually, however, they transformed themselves into a semi-hereditary group, and became a great danger to the tzar's authority until exterminated by Peter I, who built a conscript army. This army, however, consisted of men conscripted for life, and therefore had the character of a professional body. The M.P.R. remained low until the nineteenth century, and social inequalities remained extreme. Nevertheless, the fact that soldiers were now drawn from the lower strata brought into the social system an element of instability which revealed itself in the near-success of the rebellion of Stenka Razin.

The consequences of the extension of the M.P.R., which took place in the second half of the nineteenth century, were very far-reaching; and, as will be seen later, this extension was the condition of the success of the levelling revolution which broke out in 1917.

Spain

Before the Roman conquest the Iberic peninsula was occupied by a variety of peoples. They differed in many features but here, as elsewhere among societies of a similar level of complexity, we can discern a dividing line between unstratified tribes, without specialized strata of warriors, and compound societies where the warrior nobility ruled the peasants.

During the Roman domination the circumstances in Spain were moulded by the Roman system, and therefore need not be described separately. Then came the visigothic conquest which produced a semi-feudal society, where the conquerors became the warrior nobility, retaining the profession of arms for themselves.

When the Arabs, or rather the Berbers under the Arab leadership, subjugated Spain they, in turn, established themselves as the ruling warrior nobility. Gradually, with the process of ethnic amalgamation, and partly in consequence of increasing centralization of authority, the nobility lost its martial character, and the armed forces came to consist of mercenaries. This situation persisted in the emirates which came into existence after the break up of the western Caliphate.

The remnants of the Christian Spain organized themselves into several small principalities. Their stratification had been levelled considerably in comparison with the Visigothis times, owing primarily to the necessity of arming the whole populations. Serfdom has almost disappeared and, as far as social inequalities are concerned, we might say that the Pyrrenean kingdoms resembled more Sweden or Norway than western Europe or Moorish Spain. The reconquest changed this. It gave into the hands of the nobles large domains, the inhabitants of which became their serfs. Towards the end of the reconquest the M.P.R. dropped approximately to the level at which it was in other countries of western Europe while social inequalities became correspondingly accentuated. At the close of the Middle Ages Spain did not differ appreciably from other countries of Europe on this point. Thereafter, the stratification of the Spanish society became even steeper than in other countries west of the Rhine.

Until the nineteenth century the M.P.R. remained decidedly low in the Spanish kingdom. Owing to the influx of gold from the overseas possessions the kings were able to carry out a thorough centralization, of which a very important aspect was the replacement of feudal levies by professional soldiers. This, however, affected social inequalities relatively little, particularly, as in the Spanish army even ordinary soldiers were recruited mostly from among the nobility.

The introduction of universal conscription in the nineteenth century inaugurated the age of instability and revolutions.

England

The social organization of Roman Britain was similar to that of the other Roman provinces, except that the patches which remained beyond the imperial control, and preserved their tribal organization, were much larger than in Gaul or Spain. Within the area ruled effectively by the Romans the most important social unit was the villa—that is to say, the large estate worked by slaves. The armed forces consisted of the legions, by that time strictly professional.

The civilization of Roman Britain was swept away by the Anglo-Saxon invaders, who also exterminated or drove out the Celtic population. The Anglo-Saxons lived in village communities of free peasants. There were the earls, the nobility of birth; though it is not really exact to call them the nobility—they constituted no caste: they rather resembled the chiefs of the Bantu tribes of South Africa. The absence of profound social inequalities was undoubtedly connected with the military organization of these tribes, whose armed forces consisted of peasants bringing their own arms. This militia was called the fyrd, and the bulk of it consisted of foot-soldiers.

Under the impact of the Norsemen's raids the military organization was altered. The heavily armed horsemen were found much more effective than the fyrd. Naturally, ordinary peasants could not afford the equipment of a heavily armed horseman. So, a professional force became indispensable. It consisted of thegns: the nobility of service which grew out of king's retainers. They were endowed with land, and rapidly developed into a privileged stratum. Finally, the society came to be divided into the warriors and the peasants who toiled for them. Even before the Norman conquest England was covered by manors, where the serfs lived under the jurisdiction of their lords.

The Norman conquest sharpened social inequalities for a time, as conquests usually do. Nevertheless, by the beginning of the Hundred Years' War, English society had become less steeply stratified than the French, which by then had been ethnically homogeneous for several centuries. For the explanation of this fact we must look again to military factors.

During their wars against the Welshmen the English warriors learned to use the long bow. This was a most formidable weapon, far superior to any other type of bow known in Europe and more effective than the cross-bow. By combining the foot archers, drawn from among the free peasants (yeomen), with the cavalry composed of the nobles, the English leaders dealt during the so-called Hundred Years' War a series of shattering blows to the feudal hosts of the French kingdom, the core of which was the best cavalry in Europe. The English military organization was reflected in, and reflected, the social stratification. There were, of course, serfs in the England of that era, but the yeomen who provided the archers were free, and owners of the soil they tilled. Unlike their counterparts in France and Germany the English peasants were not helpless against their lords as was proved by the success of the Great Peasant Rebellion, until it was suppressed with the aid of a ruse. It surely was not accidental that the only peasant revolt in England which succeeded took place at the time of the predominance of the long bow. One is reminded of the role which the reflex bow played in China.

Significantly, when the period of enclosures begins, when more and more severe laws are being passed against the vagrants (who in reality were only peasants expropriated by their lords), the yeomen archers are needed no more; nor are they feared. The defence of the country is now assured by the navy and the mercenary soldiers, using fire-arms against which the bows would be of no avail.

The Balkans

In the Balkans the connection between M.P.R. and stratification can be seen better if we consider the matter geographically than if we try to trace variations in time. The line which divides—or at least divided until the present century—the peoples of the Balkans into two different types is the frontier between the highlands and the lowlands. The lowlands were populated since the dawn of history by the serfs and their masters. The masters changed often but the serfs underwent no substantial ethnic modification since the Slavonic invasions. Even before the Roman conquest Illyria, or rather the lowland part of it, had a reputation of being a country of humble serfs and proud warrior nobility. During the Roman domination it was sprinkled with villas, and did not differ markedly from Dacia or other parts of the Empire in what concerns the stratification. The Slavonic invasions destroyed the Roman civilization and almost exterminated the Romanized population, of which only those who sought refuge in the mountains survived. These invasions reached even the Peloponnese, and profoundly altered the racial composition of the population of Greece.

These invasions produced a wholesale regression to tribal organization; yet, the population was still stratified because, though the Slavonic tribes were fairly egalitarian, they were under the domination of the Avars—a league of Eurasian horse-nomads.

So, the stratification of the Roman villa was replaced by that of a primitive conquest state, loosely held together. When the empire of the Avars collapsed, most of its territory fell to the next wave of the invaders from the Eurasian steppe, the Bulgars; the north-eastern part of it was won by Charlemagne, and the Byzantines eventually reconquered the southern part. Everywhere in the Balkan lowlands the peasantry was dominated during this epoch by the foreign nobility. By the beginning of the eleventh century the Byzantines had already destroyed the Bulgarian state, and subjugated almost the whole peninsula, while Panonia (present day Hungary) was conquered by the Madyars and Dacia (present day Rumania) by the Petchnegs—both newly arrived from the steppe. As far as stratification is concerned, there was no substantial change. The situation remained essentially the same even after the erection of the national kingdoms of the Serbs and the Bulgarians. (The latter were by now a Slavonic nation which inherited the name of its former Asiatic masters, who dissolved themselves in the Slavonic mass.) As before the warrior nobility ruled the serfs; the only difference was that now both spoke the same language. But even this did not last very long: by the end of the fifteenth century the Balkans were under the Turkish domination, the Bulgarian and Serbian nobility exterminated or degraded. Henceforth until quite recent times the Turkish warrior nobility ruled their despised serfs with an iron hand.

All this was happening in the lowlands. In the highlands of Serbia, Bosnia and Montenegro the patriarchal tribes preserved their independence—at least *de facto* if not *de jure*—until the present century. No nobles ruled the semi-nomadic mountaineers, living under the tribal democracy tempered by patriarchy. Even in the midst of the seignorial domains, the Aromuns maintained their independent and pastoral way of life on the Carpathian peaks. And it must be remembered that among all those mountaineers every man was a warrior: a warrior above anything else.

Early Modern Europe

The supremacy of the heavily armed knight was destroyed by the improvement of infantry tactics and fire-arms; and the feudal levies were replaced by mercenary armies. This change produced far-reaching transformations of political structures, which will be discussed later, but as far as the span of social inequalities is concerned it was of no great consequence. The masses remained unarmed and excluded from military service, and therefore deprived of the levers of power which wielding arms usually gives. Serfdom, it is true, disappeared from western Europe, but that was due to economic, not military causes: the spread of monetary circulation made payments in money appear more attractive than dues in labour. Also the growth of the royal power—a complex process due to both economic and military factors—enforced the abolition of private jurisdiction, which was the corner-stone of serfdom. But it must be remembered that a real improvement does not necessarily follow legal emancipation. The peasants seldom became owners of the land they tilled: usually they were turned into tenants whose position was sometimes worse than that of serfs. They also had to bear all the fiscal burdens from which other estates were exempted. The place of the feudal lords was now taken by the courtiers, landowners and royal officers, but the privileged position of the aristocracy remained unchallenged, until the French Revolution.

The causes of the French Revolution were manifold: the thwarted aspirations of the rising bourgeoisie, the spread of ideas advocating a 'rational' social order, the feeble-mindedness of the king, the complete incompetence of his ministers, the arrogance of the aristocracy, whose intransigent insistence on their privileges was only rivalled by their indolence, the discontent seething among the poor because of the severe economic crisis—to mention only the most important. But from our present point of view the most important is the fact that a large part of the troops fraternized with the rebels and turned against the government and the aristocracy. This could only happen because the soldiers were now recruited from among the paupers. Such defection could never have occurred if the nobles themselves constituted the core of the armed forces as they did in the Middle Ages. The court nobility, moreover, relinquished even the controlling functions in the army, being satisfied with sinecures. The actual commanding was done by subalterns originating from the poor provincial nobility and even bourgeoisie. The high nobility, by letting the military power slip out of their hands, permitted a situation to arise in which their downfall was possible.

Modern Europe

The French revolutionary armies were the first in Europe to be based on universal conscription, with the partial exception of Sweden. No doubt, the revolutionary and patriotic fervour awakened among the people made the idea of 'levée en masse' appear workable and attractive. Nevertheless the M.P.R. could not have been raised so much if the technical conditions did not permit it. The most important technical circumstances involved were the improvements in the manufacture of fire-arms and the better methods of commanding large numbers of men on the battlefield. The cost of the outfit of a Napoleonic soldier was, even if we count artillery and engineers, on the average far lower than the cost of the equipment of a medieval knight. The skills necessary for a fighting-man became much easier to acquire: shooting with a rifle is much easier than archery or swordsmanship.

There was nothing in the inner structure of the European states of the nineteenth century which compelled them to adopt universal conscription. Quite the contrary: the resort to this method of raising armies undermined the position of the monarchs and aristocracies. Metternich, the chief upholder of the traditional order realized this perfectly; at the Congress of Vienna he insisted on the abolition of conscription. All European monarchies introduced universal conscription only after decisive defeats. Prussia had recourse to this expedient only after its professional army was shattered by Napoleon. The Tsars of Russia initiated reforms in this sense only after the sad experience of the Crimean War. Franz Joseph of Austria was converted to the new idea only after his troops suffered a signal defeat at the hands of Bismark's conscripts. The French Restoration Monarchy abandoned universal conscription in favour of the professional army. The July Monarchy, though more concerned about the interests of the bourgeoisie than of the aristocracy, was too afraid of the lower classes to arm them. The same can be said about the Bonapartist regime which followed. The Third Republic which succeeded it in consequence of the crushing victory of the Prussian conscript army over Napoleon's professional soldiers, always relied on conscription.

The advent of mass armies produced a new situation. The loyalty of the lower classes had to be strengthened by extending to them various rights. In Prussia, and later in Germany, this policy was perhaps most deliberate. Serfdom was abolished concurrently with the military reforms of Stein, and peasants were granted the ownership of the land they cultivated. Later, insurance schemes for the wage-earners were sponsored by Bismarck, and in this respect Germany was far in advance of other countries. In Russia the reforms of Alexander II abolished serfdom and transformed the army from a professional into a conscript force. The old professionals were not always, it must be added for the sake of precision, volunteers; they were mostly conscripts—but for life. In Austria the adoption of universal military service coincided with the reforms in virtue of which the Habsburg Monarchy became constitutional instead of absolute.

Not only was it necessary to keep the lower classes fairly contented if the conscripts were to be willing to fight, but, further, these armies were unsuitable for suppressing popular revolts. Where the ruling aristocracy was unwilling to make concessions, as in Russia, the regime collapsed. There were many peasant rebellions in Russia before the twentieth century, but they never succeeded. The 1905 revolution nearly succeeded and that of 1917 did succeed because the peasants were armed. Undoubtedly all kinds of other factors were also responsible for the success of the Russian revolution. It would either have not occurred at all or have failed if instead of a Nicolas II a Peter I had sat on the throne; or if the ruling aristocracy had been less corrupt and indolent; or if the war had been short and successful. But weak monarchs, indolent aristocracies and unsuccessful wars were not unknown in Russia before, and the regime nevertheless survived. The influence of Marxism, percolating from the West, was important in determining the subsequent turn of events; but the discontent among the peasants, which caused the dissolution of the army, was little influenced by abstract ideas; the main motives were war lassitude and land hunger. The fact remains that it was the mass army which made the success of the revolution possible.

The fate of the Habsburg Monarchy illustrates well how a state might be compelled to adopt military institutions which proved to be suicidal. Because of the rising tide of nationalisms all efforts to instil into the masses loyalty to the crown proved futile; with the result that in the hour of defeat the empire simply dissolved. Again, it was not the first major defeat which the Habsburg Monarchy suffered in its long and chequered career; but it was the first disastrous war waged with armies raised through general conscription.

The history of modern Britain does not seem to fit into the schema expounded above. There the important extensions of the franchise, i.e. some equalization of political rights, took place in the sixties and the seventies of the last century—long before universal conscription was introduced during the first World War. These reforms were followed by a number of important measures, aiming at the attenuation of the suffering of the poor; such as the legislation protecting wage-earners, industrial insurance, progressive income-tax, etc. But we should not exaggerate the effects of these steps; British society before 1914 was very steeply stratified. The rise in the standard of living of the masses was due to the general economic progress and not to equalizing redistribution. It remains true, however, that there has been a levelling tendency at work in British society ever since the middle of the last century. Among the circumstances which facilitated this trend two were of decisive importance: one was the lessening of the tension between the strata, owing to the general rise in the standard of living, which in turn was due to commercial and industrial expansion and the existence of outlets for emigration; the other was the un-military character of the ruling class. Being sheltered by the navy, Britain could dispense with a strong army; and generals did not play prominent roles in politics. The unwarlike merchants and industrialists were not able to keep down the constantly growing proletariat as the knights kept down the serfs. This could have been done, but then they would have had to cede their place at the top of the social ladder to generals and ministers of police. They preferred to make concessions to the lower classes.

It is nevertheless significant that the two world wars, fought with conscript armies, strengthened immensely these levelling tendencies. The end of the first saw the introduction of universal adult suffrage; the second brought to power the Labour Party, with its programme of 'soaking the rich'.

The contemporary situation is characterized by two opposite tendencies. The means of controlling the people and suppressing the opposition, suppression facility—as I called it, have greatly improved during the last century. Radio, telephone, photographs, finger-prints, automatic carding, talking drugs, etc. make an individual outside the ruling group helpless. So long as the government retains the loyalty of the armed forces, no revolt can succeed. These circumstances are conducive to greater inequal-ities between the ruling group and the rest of the population. This tendency has so far been balanced by the indispensability of the enthusiastic participation of all subjects in the war effort, in which now even civilians are involved. This is why even the govern-ments which have no compunction about using terror nevertheless practice what H. D. Lasswell calls mystic democracy.[1] The sub-jects are subjected to persistent propaganda that the state is really

[1] This explains undoubtedly the constant use of lies in contemporary political propa-ganda. A despotism which does not seek to dissimulate its nature does not need to be averse to truth. Frank despotisms do not need to throttle intellectual freedom in the

theirs, that the government's only care is to further their interests. Not a trace of the haughty disdain for the rabble, so frequently expressed by the aristocratic rulers of old Europe, can be found in the pronouncements of the modern despots. Prisons and concentration camps are filled, not with those guilty of '*lèse-majesté*' but with the enemies of the People. Voting at the elections, which confirms an American or a Swede in the belief that the state is really his, can easily be replaced by the knowledge that the helm of the state is being wielded by the Leader, the Saviour with whom one is in a mystic communion, or the Great Comrade, the infinitely wise Father who always knows best. Those rare fellows who insist on their right to proclaim the results of their speculation whatever they may be, can easily be liquidated or silenced. Would-be competitors of the ruling group can similarly be dealt with. But all this affects the masses very little. The sense of power which one can enjoy when dropping a vote into the ballot-box is rather negligible, compared with one's vicarious enjoyment of power through identification with the powerful dictator or the victorious nation.

Wealth and prestige are certainly more desired by the great majority of human beings than freedom of thought or the vote; and curiously, even the governments which ruthlessly suppress all opposition try to satisfy these desires of the man-in-the-street—or at least promise to do so.

There is no need to give proofs that the Soviet regime is egalitarian. Rather a correction of this view is needed. Although it undoubtedly is so in comparison with the Tzarist rule, social inequalities in Russia are much greater than is commonly imagined in the West and the present trend is towards their accentu-

same way as does Stalinism. Even social sciences flourished under the Romanovs, Habsburgs and Hohenzollerns. An unabashed autocracy may allow considerable freedom of expression except when it leads to a direct attack on the ruler or his right to rule. Indeed, despots often fostered free thought and even protected daring thinkers from the wrath of a bigoted populace. Examples illustrating this point are numerous; it is enough to mention Frederick II Hohenstaufen, Akbar, the Ptolemies, the Medicis, al Mamun. This is nothing abnormal because science, so long as it remains non-evaluative, can no more prove or disprove the right to rule of a hereditary monarch than it can prove or disprove the right of a majority. Recognition of such norms is the matter of ethics, unsusceptible to logical proof. Modern totalitarian regimes are inimical to free thought not so much because they are autocratic as because they rest upon a Big Lie.

ation. On the other hand, the prevalent ideas about the effects of Nazi rule on the social stratification of Germany are quite wrong. Hostile propaganda pictured Hitler's regime as a ruthless exploitation of the worker in the interests of a small clique. In reality Hitler bestowed upon the German lower classes considerable economic benefits. He took away, it is true, all their political rights and suppressed mercilessly all opposition, but he also built houses for them, provided them with all sorts of amenities and abolished the constant threat of unemployment which hitherto hung over their heads. The power of capitalists and landlords was curtailed. An owner of a factory, for example, had no right to dismiss an employee without official sanction. Later, the people were asked to work for 'guns instead of butter'; but this, they were made to believe, was a necessary step in their career of conquest, the fruits of which would be wealth and power for every German. This is the explanation of the passionate loyalty which the German people felt for their Führer. Nobody can doubt it who has seen lower-class German women fall into ecstasy when hearing his voice, or witnessed the stubbornness with which the German soldiers fought to the bitter end. The German people resembled a war-band devoted to their leader, whose lust for power was matched by their eagerness for booty.

It is significant that the regime which more than any other totalitarian dictatorship was a mere cloak for pre-existing privileges, and was the least egalitarian—Italian Fascism—proved to be the weakest and collapsed ingloriously.

The foregoing evidence is sufficient, I think, to prove that the M.P.R. is one of the strongest determinants of stratification; and that, on the other hand, stratification tends to reflect itself in the M.P.R. In other words, the height of stratification tends to co-vary with the M.P.R.

*　　　*　　　*

It is extremely probable that the M.P.R. exerts very strong influence on the position of women, which is usually higher in societies where they participate in wars, even if in a subsidiary role, than in societies where they do not. This proposition could only be proved by a special investigation which I cannot undertake, but a cursory glance over the evidence definitely points this way. In Europe, women made the greatest strides towards equality during and after the two World Wars, in which they proved to be very useful. Among primitive tribes where women help in wars their standing is high. On the other hand, in societies where their status is at its nadir they are absolutely precluded from any contact with things military. There are, of course, many other very important factors which influence the status of women: e.g. magical and religious beliefs.

I must add finally that the variations in M.P.R. due to changes in age composition do not affect stratification.

III

The Size of Political Units
and their Cohesion

ATTACK VERSUS DEFENCE

IN every political structure there are at work centripetal and centrifugal forces, promoting territorial cencentration or dispersion of political power respectively. Military factors affect their balance, thus largely determining the territorial distribution of political power.

In order to avoid misunderstanding I must point out that territorial cencentration (or dispersion) is only one form of concentration (or dispersion) of power. When, for instance, towns become more independent of the capital, that is territorial dispersion of power, but when the judicature emancipates itself from control by the executive that is non-territorial dispersion of power—in this case functional. Another form of non-territorial distribution of power is the competitive (or better, substitutive), like that existing between political parties or churches thriving within the same territory.

The first proposition which I shall seek to establish concerning the relationship between military factors and territorial distribution of political power, is the following: other things being equal, the predominance of attack over defence promotes the territorial concentration of political power (centralization), while if defence becomes the stronger form of warfare, a trend towards the territorial dispersion of political power (decentralization) is likely to ensue.

In view of the difficulty, frequently met, of deciding when the territorial concentration of political power is to be called centralization and when territorial expansion (should the process which went on in France from 1400 till 1700, for instance, be called the centralization of France or the expansion of the Capetian patrimonial state?), we might restate the above proposition as follows: other things being equal, the predominance of attack over defence tends to diminish the number of independent governments within a given area and to widen the areas under their control, and/or facilitates the tightening of control over the areas already under their domination; while the superiority of defence tends to produce opposite results.

We should expect, therefore, that when the art of fortification is more than a match for the existing siege-craft, the size of the political units will be smaller, and their number within a given area greater, than in times when every stronghold can easily be stormed. There are, of course, disturbing factors, but historical evidence supports this proposition.

The formation of the Chinese Empire was connected with the changes in armament and tactics which preceded this event. During the era of small virtually independent principalities, feudal chariotry was the main arm. Its attacks could not be very effective because any ditch could stop it. Feudal lords resided in strongholds which were well-nigh impregnable. This situation was changed during the so-called period of warring kingdoms, which ended with the conquest by the Ch'in of all rival principalities. The cavalry, attacks of which were much more difficult to repel, replaced the chariotry, thus making battles decisive. At the same time there appeared siege-engines with the aid of which even the strongest fortifications could be stormed: battering-rams, catapults, rolling towers, etc. The balance between attack and defence clearly changed in favour of the former.

Seven centuries earlier a similar development provided the basis for the foundation of the Assyrian Empire, far greater than any hitherto seen. Assyrians were the first to develop the effective technique of assaulting strongholds by means of battering-rams and other siege-engines. They also created the first cavalry.

The welding of Greek cities into one empire had also something to do with the balance between attack and defence. Philip of Macedon introduced into his armies the artillery of catapults and balistas, and other siege-engines. Until then the Greeks knew no other means of reducing fortified places than starving them out.

Not the least important among the circumstances which brought about the Roman conquest of the Mediterranean world, was the fact that the Romans brought siege-craft to a perfection unheard of before. As no improvements in the art of fortification took place at the same time, the balance was tipped in favour of attack. The Roman siege-craft was really impressive and required considerable technical knowledge on the part of the officers and even the ordinary soldiers.

The disintegration of the Roman Empire in the west was accompanied by a change in the balance in favour of defence. To be sure, this circumstance played no part in the decline of the Roman Empire but was its result. The Germanic hosts which replaced the legions were too uneducated, unruly and unaccustomed to methodical work, to be able to perpetuate the Roman siege-craft. But once this change in the balance between the two forms of warfare occurred, it hampered all attempts to restore the unified empire. When the art of storming fortifications partially revived in western Europe, its resuscitation came too late to undermine the strength of the baronial castle, which had come into existence in the meantime. It was a far more formidable stronghold than a fortified city because its site was chosen purely for its natural advantages, whereas the walls of a town had to follow its contours. Containing no inhabitants except its garrison it could withstand a blockade for a very long time. The castle was the main basis of the independence of the barons.

In central Europe, where the Roman siege-craft was never known, the balance was tipped in favour of defence by improvements in building technique, which substituted walls of masonry for banks of earth and palisades. By the time of the Saxon Kings proper-castles had begun to appear. From then onwards the power of the Kings was steadily reduced and appropriated by feudal magnates.

It is highly significant that siege-craft remained on the former high level in the portion of the Roman empire which preserved itself intact—Byzantium.

Towards the end of the Middle Ages national monarchies began to rise in western Europe, and in Germany the small barons were gradually subjugated by the princes. The causes of this process of territorial concentration of political power were multifarious: the revival of monetary economy, the improvement of administrative technique, the growth of towns—to mention only the most obvious. But military factors were at least equally important. It was the cannon, which came into use at this time, that smashed the main pillar of feudalism: the baronial castle, hitherto impregnable.

From the epoch of Renaissance to the nineteenth century, the oscillations in the balance between attack and defence were too small to produce far-reaching effects in the political sphere. The indecisiveness of eighteenth-century wars was due not to this but to other factors. They were more in the nature of royal games than life-and-death struggles between the nations. The Napoleonic wars brought a change: the aim of war came to be the destruction of the enemy. Fortresses could still resist for very long periods but armies learnt to by-pass them. The situation, however, was not radically different from what it was during the campaigns of Gustaf Adolf.

In the first World War the machine gun forced trench warfare upon the armies. The preponderance of defence became indeed startling. It would be too much to say that the sudden multiplication of states in Europe was the consequence of this new military situation. Nevertheless, after due allowance is made for the influence of nationalism and other factors, it seems probable that many of the new states would not have been created if they had been thought to be completely defenceless.

In the second World War attack proved to be supreme for reasons too well known to be discussed. And again a coincidence must be noticed: the number of really independant states in Europe has diminished. Eastern Europe is virtually one state governed from Moscow. In western Europe the number of states has not diminished but they are independent only on sufferance. They could easily have been made into American dependencies. Here again we see that military factors are not the only determinants of the political situation; they open certain possibilities but these can remain unused. Nevertheless, a certain measure of military coalescence of western European nations, dictated by the impossibility of individual defence, has taken place since the end of the war.

The latest weapons have rendered any defence, except retaliation almost impossible, and have made the military conquest of the whole world quite feasible.

THE SIZE OF POLITICAL UNITS

THE INFLUENCE OF FACILITIES OF TRANSPORT
AND COMMUNICATION

A contest between mobile armies is more likely to issue in a clear-cut decision, while a draw is more likely if the armies are slow-moving. This is clear if we compare, let us say, battles in which the Mongol cavalry was involved, with clashes between European feudal hosts. Good transport facilities permit the utilization of the resources of wider area. Also, with mobile armies rebellions can be swiftly suppressed. The rapidity of communications enables the central government to supervise local officials effectively, thus preventing the disintegration of the state. The radius of efficient military action also depends on the state of transport and communications. The same factor together with the technique of commanding, sets the upper limit to the size of armies which can be directed strategically or tactically. With the improvement of the means of transport and communications, big states, which hitherto derived little advantage from their size, become relatively stronger, and can subjugate their smaller neighbours more easily.

As a consequence of these facts we should expect, that whenever an improvement in the technique of transport and communication occurs, which is not counterbalanced by an increase in the weight of equipment nor by a rise in the relative effectiveness of defence, the size of the political units will tend to grow. Let us see if history bears witness to this contention.

The earliest traces of the use of horses in Mesopotamia are contemporaneous with the foundation of the first extensive empire in that land. The Hyksos who spread the use of horses into Syria and Egypt created an empire wider than any hitherto seen. After their expulsion from Egypt, the Egyptians, now possessing horses, embarked for the first time upon conquests abroad. The Assyrians, who created an empire greater than any that existed previously, used chariotry more extensively than any other people. They increased, moreover, the mobility of their armies by forming the first cavalry we know of. Horses it must be noted were until then used for drawing chariots, not for riding. An even greater empire was built by the Persians, who were renowned as horsemen, having been originally pastoral nomads. In India the rise of the great kingdoms, which swallowed a multitude of small independent

79

principalities, coincided with the introduction of horse-riding and the domestication of elephants, which improved transport tremendously. China was welded into one empire soon after horse-riding replaced chariot driving. It was because of their great mobility that pastoral nomads, like the Mongols, Arabs or Turks, were able to establish their gigantic empires. It must be noted, however, that in their pristine state, i.e. before the conquest of sedentary populations, the nomads never created any states, because no real control can be exercised over tribes continually on the move. The Roman Empire, based on maritime transport, could not have arisen without great improvements in ship-building and navigation previously effected by the Greeks. Similarly, the introduction of the compass and improvements in the art of sailing enabled the Europeans to build their colonial empires.

In view of the foregoing evidence, we should expect that the unprecedented progress in the technique of transport and communication, which occurred in the nineteenth century and after, would lead to further political unification; but it is by no means obvious that it has. Since the beginning of this century several great empires have disintegrated into a number of independent states. There are apparently forces at work which counteract the influence of the factor here discussed; I shall deal with them in the following chapters. Even though, however, no definite trend towards the diminution of the number of states during this period is discernible, within each political unit centralization has proceeded at a very rapid rate.

The relation between political unification and centralization and the road system is very striking. Road-building has always been one of the chief preoccupations of empire-builders, like the Persians, Romans, Incas, and many others down to modern times. The dissolution of empires is often followed by the decay of the roads, which cannot be maintained without an efficient centralized government. In Europe roads were completely neglected between the fall of the Roman Empire and the rise of the modern national monarchies.

The conclusion of the foregoing is that, other things being equal, improvements in the facilities of transport and communication foster territorially centripetal tendencies of political structures, while their deterioration strengthens the centrifugal tendencies.

ARMS OR ORGANIZATION

Centrifugal tendencies will be strengthened if the dominance of the warrior stratum is based on individual superiority; when it is due, for instance, to the possession of arms better than those which their subjects can procure. In such a situation they may rule individually, each governing a number of subjects, or serfs, or servants of some kind, as was the case in medieval Europe or in South Africa of the last century. If, on the other hand, the preponderance of a group of warriors is due to their superior, closely knit organization, then they can preserve their dominance only if they maintain their cohesion. They must, therefore, rule collectively not individually. The Spartan state and the Inca empire present classic examples of such collective dominance. The arms with which the Inca tribe conquered their empire were of the simplest kind: clubs and shields, which were not greatly superior to weapons which could be made by the surrounding tribesmen. Inca victories were due to their better organization and above all to the iron discipline unknown to their neighbours, which could not be relaxed if dominance was to be maintained. This situation produced one of the most rigidly disciplinarian states in the world.

MONOPOLY OF SUPERIOR TECHNIQUE

Many empires have been created because the exclusive possession of superior armament or tactics enabled some peoples to subjugate others. This monopoly was an essential condition of the existence of these empires, and its loss usually led to their disintegration.

Examples of conquerors owing their successes to onesided superiority of tactics or armament abound. The Zulus succeeded in extending their sway over a large part of Southern Africa because of the military innovations of their king Chaka, who replaced the disorderly tribal host by well-organized regiments. He also taught the Zulu warriors to fight in close order instead of the traditional loose grouping, and to use their throwing spears for thrusting. The exclusive possession of iron underlay the triumphs of the Assyrians and of the Ch'in state in China. The conquests effected by the Persians, Macedonians, Romans, and many other peoples, were preceded by military innovations which enabled them to attain preponderance. The successes of the Napoleonic and the Nazi armies were the reward for developing new methods. The same circumstance accounts for the colonial expansion of the European nations in modern times.

The monopoly of a superior technique may disappear quickly, as in the case of the Napoleonic and the Nazi armies, or it may be maintained for centuries, l,ke the European preponderance over the peoples of other continents. The superiority due to some relatively minor innovation is usually transient; if it is based on some radical differences of culture it must be more enduring. It is, for instance, much easier to imitate new tactics than to adopt industrial economy. Superiority may also be long-lived if it is rooted in the national character, like the Roman discipline. The Roman tactics could not be easily imitated by the Gauls or the Teutons because they could not adapt themselves to the necessary discipline. Moreover, among ancient and primitive peoples methods of warfare are often deeply embedded in magico-religious beliefs and therefore considered sacred and unalterable; so that monopoly of better tactics can be retained for very long periods.

The effects of a loss of this monopoly are well illustrated by the fate of the Roman Empire, to take one of the best known examples. In Caesar's time the main weapons of the Teutonic warriors were wooden clubs and wooden spears with iron points. At the time of the Great Invasions we see them armed with iron swords and axes, protected by great shields, helmets, and even light body-armour. It is obvious that this military progress, which was the result of the general technical progress of these peoples, must have contributed to the fall of the Roman Empire. The European domination over the Asiatic peoples became increasingly difficult as these peoples learned more and more about Western technology and methods of organization. After the second World War the task of keeping them down became too heavy for the exhausted colonial powers; though it must be remembered that ideological factors, namely, the spread of nationalism in Asia and of pacifism and humanitarianism in western Europe, also played an important role in undermining the colonial rule.

It must be noted that the cases of the attainment of a monopoly of superior military technique leading to the foundation of empires, are much more numerous than the cases of its loss producing their disintegration. The explanation of this is that conquered peoples are usually disarmed and thus made helpless. Conquest states generally disintegrate because the conquering group loses its cohesion, or they are overthrown from without.

Modern military technique produced two contrary effects. On the one hand, it strengthened the centripetal forces, by making subjugation of distant regions easier; but on the other hand, it fostered the disintegration of multi-national empires, because universal conscription became an unavoidable condition of military strength, and armies raised in this way were of little value unless permeated by patriotism. The arming of heterogeneous elements, moreover, enabled them to assert their nationalistic claims with greater vigour; and once the discipline of the army was broken in consequence of defeat, the same arms were used in national insurrections. This was the main cause of the collapse of the Habsburg and Ottoman Empires. The position in India was similar. India could not be defended without large indignousarmies; but such armies could not be trusted to be loyal to the British Crown. It is not accidental that the most emphatic promises of independence were given to the Indians during the two World Wars.

It is certain, however, that the consequences of the introduction of extensive conscription in ethnically heterogeneous states would be less pronounced had this not been the age of nationalisms.

One of the main difficulties of all rulers of non-industrial societies is to prevent the local officials from becoming independent. The outcome of these efforts depends largely on the method of remunerating soldiers and officials. These can be remunerated in a number of possible ways, which can be combined: (1) they may receive only goods from stores under the control of the central government, like the troops of the Old Kingdom of Ancient Egypt; (2) they may receive money only, like the armies of early modern Europe; (3) they may be rewarded with land, and this in two forms: (a) they may be given land which they will cultivate

83

themselves, while being exempt from various burdens imposed on ordinary peasants, like the warriors of the New Kingdom of Ancient Egypt or the Russian cossacks; (*b*) they may be given the right to perceive dues from the actual cultivators of the land, like the Spahis of the Ottoman Empire; (4) they may be given the right to collect the revenue and be required to surrender a part of it to the central government, like the Chinese mandarins.

The crucial point is whether the local administrators and military officers depend for their income on the central government or not. It obviously is very difficult for the central government to exercise effective control over officials who do not have to wait for their pay but can help themselves, perhaps even without informing the central government about the size of their incomes. Moreover, in such a situation the fiscal authority is combined with the administrative and perhaps the military; this is contrary to the principle of separation of local spheres of authority which is the cornerstone of all centralization.

Centralized remuneration is greatly facilitated by the existence of money, which permits the drawing of all incomes into the treasury of the central government, whence all payments are made. Special organs for collecting taxes can be created. The provincial officers, being accustomed to expect their rewards from the central government, are inclined to remain obedient. Centralized remuneration in kind is more difficult. Transportation of bulky goods became easy only with the advent of railways and steamships, and for this reason, until the industrial era, a centralized state based on this principle of remuneration could only exist in river valleys, like those of Egypt or Mesopotamia. The Inca Empire is the only seeming exception. But the degree of centralization prevailing in that state has been greatly exaggerated: the ancestral village community was not greatly altered by the Inca conquest. Moreover, the Incas, being recent conquerors, had to hold together to maintain their position. The economic factor in question, it must not be forgotten, is only one of the many which determine oscillations in the territorial distribution of political power.

Centrally directed remuneration makes the existence of centralized administration possible; but, on the other hand, since this bureacratically organized administration is indispensable for performing the variegated functions involved in collecting taxes, administering stores, keeping records etc., any circumstances which hinder the development or functioning of bureaucratic administration compel the government to have recourse to methods of remuneration which strengthen centrifugal forces. Among such circumstances the most obvious are illiteracy and the lack of knowledge of the administrative technique. Progress in these techniques naturally facilitates centralization. The best example of the consequence of regression in this field is provided by the disintegration of the Roman Empire. The Germanic conquerors, being illiterate, could not work the complicated machine of imperial administration. They could and did use the Romans as clerks, in the same way as the Mongols or the Manchus used the Chinese literati or the Arabs used the Persian and Syrian scribes, but this class of educated Romans soon disappeared. The tremendous power enjoyed by the Church during the Middle Ages was due, not least, to its monopoly of literacy and education. This power began to wane when educated laymen appeared.

When we hear about money becoming scarce in the Late Roman Empire it does not means that the actual quantity of coins diminished. On the contrary, until the complete disintegration of the empire it definitely increased because so many emperors indulged in debasement of the coinage. Nor is there any evidence that the propensity to hoard increased. *A priori* we might expect an intensification of hoarding in times of trouble, but not if prices are rising. It was apparently the curtailment of the volume of commercial transactions which, together with the policy of debasement, produced rises in prices against which imperial edicts were of no avail.

A drop in the volume of transactions means a drop in the velocity of circulation of money. Money which does not circulate is of no importance in economic life. The monetary economy disappeared because commercial activities ceased.

Intensive trade is possible only if there are adequate means of transport, if travelling is safe, and if legal order is stable. These conditions can exist over wide areas only under fairly centralized government. Medieval merchants were staunch supporters of the kings in their struggle to subdue the feudal lords, because this was their only hope of being freed from robber barons and the profusion of tolls. Civil wars and foreign disturbances naturally undermine monetary economy. We find here a phenomenon of mutual reinforcement of two processes: initial political disturbances undermine monetary circulation, thus producing circumstances propitious to further political disintegration, which in turn further hinders trade, and so on. Such were the cases of the Roman Empire and the Arabian Califate. But, as in the case of the Parthian kingdom, the series may also start with the initial decline of trade, which will foster political disintegration, which in turn will lead to further decline of trade and so on. The most frequent cause of the decline of commercial activities, not produced by political disintegration, is the shift of trade routes or their interruption. I must stress, however, that both concentration and dispersion of political power can be due to other than economic causes, and may be unaccompanied by any substantial changes in this sphere.

In the history of the Arabian Califate the connection in question can be very clearly seen. The Arab conquest of many hitherto independent and mutually hostile lands produced a great intensification of trade which, I may remark incidentally, had a great deal to do with the contemporaneous efflorescence of Arabic culture. The failure to establish a fixed rule of succession, together with the inability of the Califs to control their Turkish mercenaries, enabled the local governors to acquire more and more independence. The inevitable loss of tribal cohesion among the conquerors, resulting from geographical dispersion and transculturation, also contributed to this effect. Some of these local officers became feudal lords, nominally recognizing the authority of the Calif; others proclaimed themselves independent sovereigns. But all, however minute their domains, claimed the rights to levy tolls, often ruinous. Already the initial dynastic strife and troubles with mercenaries, by undermining trade and monetary circulation, compelled the Califs to assign lands to soldiers and officials, thus making control over them difficult. Now, the more independent they became, the more frequent became wars between them, the greater was the decline of trade, and the more necessary the assignment of land as a method of remuneration; and, therefore, the stronger became the centrifugal tendencies. The 'vicious' circle was complete.

The major part of the lands once ruled by the Califs of Baghdad was re-united through conquest by Seljuk Turks in the eleventh century. Then the cycle repeated itself almost exactly until it was cut short by the Mongol conquest.

The Parthian kingdom enjoyed considerable benefits from the transit trade between China and the markets of the Mediterranean. The disruption of this trade by the nomads of central Asia was followed by the disintegration of the Parthian kingdom.

The process of disintegration of the Roman Empire in the West differs in many respects from similar processes, which occurred in the medieval Near East. The manor, for example, which was so important in Europe, never developed in Islamic countries: or rather it never became a dominant institution there. Nevertheless, the connection between the disappearance of monetary economy and the political disintegration can be observed in all these cases.

Monetary circulation and political centralization further each other in their development as well as in their decline. In Europe the rise of centralized monarchies fostered trade, the expansion of which weakened feudalism. With money derived from taxes imposed upon towns the kings were able to recruit mercenary soldiers, with whose help they succeeded in subduing their unruly vassals. More and more lands could be placed under direct control of the royal officials, who, being paid by the kings, remained their obedient servants.

In the ancient Near East the Persian Empire, greater than any hitherto known and also the first to introduce the separation between civil and military governors, came into being at the same time as money was coming into use.

In China the centralizing reforms of the T'ang dynasty, the creation of a great standing army of mercenaries, the further extension of the bureaucratic machine and, later under the Sung, the separation of cival and military local authorities were made possible by the intensification of monetary circulation.

METHODS OF EQUIPMENT

A government will be able to control its troops more strictly if it not only pays them but also provides their equipment. The importance of the latter task depends primarily on the cost and the rapidity of wear of the equipment. If, as in modern armies, the armament becomes so complex that it can be neither produced, nor stored, nor manipulated by individuals, self-equipment becomes impracticable. It follows, therefore, that the increasing complexity of armament favours political centralization. Thus, for instance, the advent of artillery had this effect. Cannons could not be produced by every robber baron, but only by princes who ruled towns populated by skilled craftsmen. The contemporary situation is so obvious in this respect that no comments are required.

In non-industrial societies armaments can as a rule be produced in small workshops, and the monopolization of its supply by governments is the expression of their desire to assert their control and not dictated by technical necessities. Usually, the rulers prefer to equip the armies, because they know that such an arrangement strengthens their authority. Whether they are able to put it into practice depends largely on the same circumstances which I discussed in connection with the problem of remuneration: efficient means of transport and well-organized bureaucracy are needed for this purpose. Apart from these factors, monopolization of the production of armament is easier if the supply of raw materials can be conveniently controlled. In Ancient Egypt copper ore had to come from outside through a narrow isthmus; it was easy for the government to see that all of it went to state workshops and none was diverted to private uses.

The change-over from self-equipping hosts of warriors to troops equipped by the government is a necessary condition of centralization. Here again we meet the relationship of mutual reinforcement between two processes. Anything that favours centralization and bureaucratization promotes the change-over from self-equipping to government-equipped armed forces; whilst anything that favours such a change also promotes centralization. The same can be said about the processes proceeding in the opposite direction.

Historical evidence supporting the generalizations formulated above is abundant, but its detailed analysis would involve a great deal of repetition of what has already been said, and in any case would be rather superfluous, as this problem has already been treated with unsurpassed mastery by Max Weber in the works indicated in the bibliography.

FRICTIONAL FACTOR

Nationalism is probably the most important among non-military factors which complicate constellations of forces determining the territorial distribution of power.

THE SIZE OF POLITICAL UNITS

Modern nationalism is very often depicted as something unique, without a precedent. There can be no doubt that this is an exaggeration; modern nationalism is a special case of ethnocentrism, which is as old as mankind. Warfare between very primitive tribes often reveals almost unbounded cruelty towards the foe, combined with equally great devotion to one's own group. Nor is resentment against foreign rule anything new. The peculiarity of modern nationalistic movements is that, unlike anti-foreign movements, in the past which were directed against actual oppression, they resent even the mere nominal allegiance to a foreign government. This characteristic is the consequence of democratic ideas, the actualization of which was greatly fostered by the introduction of universal conscription.

Ethnocentrism being an omnipresent phenomenon, the real problem is the boundary between 'we' and 'they'. This can be split into two questions: that of the comprehensiveness of group consciousness, and that of the selection of the group distinguishing mark which is considered the most important.

The size of the biggest group to which loyalty may be felt varies greatly. The loyalty of the European peasant of not so long ago did not reach beyond the horizon of his native village. It was the enlargement of his contacts with the outside world, due mainly to industrialization and education, that enabled him to conceive the idea that he belonged to a nation comprising thousands of villages like his own. As 'we' can be conceived only in contrast to 'they', the peasants became nationally conscious when he came in contact with members of other nations, or at least apprehended their existence.

The bond of nationality came to be regarded as more important than the bond of religion mainly because the sway of traditional religions was weakened. The causes of this weakening present a problem beyond the scope of this book.

The foregoing considerations provide a good example of the often forgotten truth that social forces do not always work in one direction only: that even the same process of change may have consequences which will 'push' society in two opposite directions. Thus we saw that the improvements in transport, communications and printing made possible the effective administration and subjugation of wider territories, while industrialization integrated them economically. But the same transformations fostered nationalism,

89

which proved to be the most disruptive force at work within multi-national empires.

The European rule in Asiatic and African lands had (or has) a curiously contradictory and self-defeating character. On the one hand the natives are expected to obey the foreign government and are hemmed in on all sides by the colour bar, while on the other hand, as soon as they come in contact with European literary culture, they imbibe the ideals of democracy, equality of all men and so on. The more westernized they become the more they resent European rule. Such conflicts were non-existent in empires where authoritarian ideology pervaded both the rulers and the ruled. The metropolitan peoples, on the other hand, become less and less willing to hold the colonies by sheer violence. Without going deeper into the problem of the causes of this pacifism, I may say that it could not spread if these peoples were not becoming less fertile and more prosperous, and their manners milder.

IV

Subordination and Hierarchy

IN Chapter I I tried to show how M.P.R. influences social stratification. But the structure of a society depends also on the factors which determine the inner structure of the warrior group, be it comprehensive or restricted. There are two extreme structural possibilities here: the warrior group may constitute a monocratically ruled, disciplined body, or it may be organized along the lines of independence and equality of all warriors. Any modern army may be taken as an instance of the former possibility; the latter is exemplified by the fighting men of a stateless tribe, let us say the Kavirondo Bantu. Even the last named, it is true, are not altogether devoid of chieftainship, but this is transient and very restricted.

If M.P.R. is very high, the armed forces at their full strength comprise the whole adult male (or even just adult) population. Then during a war or the period of preparedness for it, the inner structure of the armed forces is practically co-extensive with the totality of the social structure. In an extremely militant society such coalescence becomes permanent. In such a situation the factors which determine the degree of hierarchization of the armed forces also determine whether the political organization will be democratic or monocratic, or something between the two poles.

If professional warriors are the dominating stratum, the inner structure of the armed forces determines whether the state is an absolute monarchy (like the Ottoman Empire) or some other form of monocracy; or whether it is an oligarchic republic of the warriors, like Sparta.

We must, therefore, enquire into the problem—on what does the inner structure of the armed forces depend? But before we proceed to this question we must direct our attention to certain complicating factors.

Substitute disorder for war.

SUBORDINATION AND HIERARCHY

INFLUENCE OF THE INCIDENCE OF WAR

Co-ordination of efforts is a necessity in a struggle between groups. In large groups it can be achieved only on the basis of subordination and unitary command. The proposition that unitary command is advantageous in war is so evident that no elaborate proofs are required. Everybody knows that armies always have commanders and are more disciplinarian than civilian organizations. Moreover, the regime of committees, so successful elsewhere, always failed in war. Even primitive tribes which have no heads in times of peace have them in war. It is true, of course, that many other human endeavours may require co-ordination of individual actions and unity of command. Generally speaking, groups which have to cope with emergencies, like crews of ships or squads of firemen, are more monocratically organized than, let us say, trade guilds or sports clubs. But war is on the whole an emergency in which the co-ordination of actions of great numbers is more than ever imperative. The more frequent and serious, therefore, is the warfare in which any armed forces are engaged, the more disciplined they are likely to be. And, as the same factor tends to make military organization preponderant among forces shaping social structure, we should expect that, other things being equal, the more frequent and serious are the wars in which a political unit is engaged, the more monocratic will be the form of its government. Allowance, of course, must be made for the inertia of the past.

Factual evidence supporting this statement is profuse: it is so abundant indeed that its examination could fill a whole volume. Luckily, there is no need to present its full array here, because much of it can be found in Herbert Spencer's *Principles of Sociology* (particularly in Part V) and Pitirim Sorokin's *Man and Society in Calamity*, and *Social and Culture Dynamics*, vol. 3. I shall limit myself, therefore, to a few examples.

The early pre-historic cultures of Europe which were predominantly peaceful, judging by the lack of developed weapons and fortification among their material remains, left no traces of powerful chieftainship in the shape of distinguished abodes and graves, which appear simultaneously with elaborate weapons and fortifications. From this does not follow, however, that all monocratic authority is of military origin; its other root lies in the power of

92

the priest and the magician; and the earliest archaic societies, like the Sumerian city-states, were predominantly peaceful, though almost monocratically ruled. Among the ancient Slavonic and Germanic tribes the institution of kingship developed in response to the needs of war. It developed earliest in the tribes threatened by invasion, or in those which embarked upon conquests. The authority of the chiefs of conquering war bands usually atrophied when the conquerors began to feel themselves safe. The most despotic kingdom of Europe rose at its extremity, where it was exposed to attacks of Eurasian nomads: in Russia. Chieftainship is signally absent among the most peaceful of primitive peoples discovered by ethnographers.

In order to estimate the effects of war we must know not only how frequent but also how serious it is. Obviously the consequences of a war which consists of periodic forays must differ from those of a struggle for life or death. The more earnest is the struggle, the more indispensable is the unity of command. It is a very common feature of wars waged by allied states, that they elect a common commander only when the situation becomes very serious. The Romans, who in the early days of the Republic were so suspicious of any personal authority that they even had two consuls commanding armies on alternate days, nevertherless provided for the insitition of a dictator, possessing absolute authority, in times of great danger. Holland became a monarchy when it was fighting for existence. During the recent wars parliaments, usually jealous of their prerogatives, bestowed tremendous powers upon executive heads of states. Only unimportant wars in distant regions, like colonial expeditions, can be conducted by a slowly deliberating parliament, the sole indispensable unitary command being that of a commander in the field.

The habits of command and obedience generated by the needs of war tend to persist in times of peace. And naturally, if the wars are frequent and peace is rare, even peace-time political organization will resemble that necessary for waging war. Not only actual war, but even the imminence of it, foster monocracy. Evidently, preparedness means being organized in such a way that war could be waged without lengthy re-organization.

These propositions apply not only to states and tribes but to all groups. All groups engaged in intense struggle tend to be monocratically ruled, be they small gangs or giant parties. One of the

93

reasons why revolutions usually produce dictatorships is that only a well-disciplined party can succeed in a ruthless contest. The threat to a social stratum does generally induce it to obey a leader. Thus conquest states tend to be monocratic, because the conquerors can hope to maintain their position only if they remain obedient to their leader. As soon as ethnic amalgamation takes place, discipline tends to be relaxed. The barons of England began to curtail the power of the king only when they ceased to be Norman and become English; they no more felt submerged in a sea of hostile foreigners.[1]

The fact that some of the most warlike primitive tribes have no permanent chiefs seems to contradict the above statements. North American Indians, for example, most of whom must be counted among the most warlike peoples of the world, were notably democratic in their government—with the exception of the Natchez. Of many of them it can be said that they had no government at all in the sense of a body of men entitled to give orders to others. Most of them, it is true, had war chiefs and hunting chiefs, as a rule elective, but their authority was limited to these spheres. The requirements of warfare did then produce patterns of command and obedience but these patterns were not extended to peaceful activities. The question naturally arises: What circumstances promote such extension?

The ethnography of East Africa throws some interesting light on this problem. The outstanding fact in the history of this region is its conquest by hamitic cattle-herders, who drove out or subjugated Bantu agriculturists. Several sizable kingdoms arose in consequence of such subjugation. It is significant that, according to traditions still persisting among the populations of these kingdoms, neither the conquerors nor the conquered had any kings before the conquest. Moreover, the Masai, who instead of subjugating the agriculturists drove them out, never evolved a permanent chieftainship, except a paramount priest-king without executive authority. In view of the fact that the Masai are the most warlike of all hamitic peoples of East Africa, it would seem that it is not warfare in itself but the conquests to which it may lead that constitute the matrix of monocracy. One reason for this is that

---

[1] The same factor explains to a large extent why the southern states of the U.S.A. differ from their compeers by exhibiting a bent towards one party political system and the Latin American 'caudillismo'.

the importance of the co-ordinating agency increases as the body politic grows, and such growth beyond the community based on kinship is, almost without exception, effected through conquest. In the process of building a conquest state, moreover, the social organization both of the conquerors and the conquered is deranged thus leaving more scope for the creation of bonds of dependence extraneous to kinship.

From the foregoing the conclusion emerges that only in a fairly large group does warfare necessarily produce monocracy. This leads us to the problem of the influence of size.

INFLUENCE OF SIZE

The larger the group, the more important is the co-ordination of the actions of its members; particularly in an emergency. We should expect therefore, that, other things being equal, the larger the group, the more prominent should be the monocratic and hierarchic traits in its organization. It is a well-known fact, moreover, that as the size of a collectivity increases it can be dominated by a proportionately smaller minority; one policeman can hardly keep down 100 civilians but 1,000 policemen can easily keep down 100,000 civilians. This is because the advantage derived from being organized grows more than proportionately to the numbers involved. There is, moreover, another factor of great moment which has been analysed with great perspicacity by N. S. Timasheff. I cannot do better than quote from his *Introduction to the Sociology of Law* (Cambridge, Mass., 1939, p. 186):

'The difficulty of transforming subjective attitude of dissatisfaction into a social force strong enough to overthrow a power structure grows with the dimensions of the system. In a small system ... it is obvious to every member that the power relationship consists merely of the dispositions of the partners but even within a group of two ... there is a certain tendency to see not two, but three elements, viz. the dominator, the dominated and the power relation itself. Such an idea is often unconscious and, of course, not very permanent, since an effort on the part of the subject or negligence on the part of the dominator, such as his failure to use his power, are sometimes sufficient to destroy the illusion of this third element of the objective relation uniting the two, and even to destroy the power relationship itself.

'As has been observed by Simmel . . . the situation changes completely when the group is increased and consists no longer of two, but of three members. In a system of three elements, A, B, C, there are three links, AB, AC, and BC, and for every member one of the links is independent of his will, is 'objective' . . . With every new member the number of links which are objective for each one increases very fast. . . . Now almost every 'objective' link means . . . . a reinforcement of the necessity of submission, because in every group-member the submission reflexes are reinforced by the submissive attitudes of the others. . . . A power system is not a mere sum of the submissive attitudes of the subjects. . . . The sentiments . . . of others are objective facts. One has to submit even if one is disgusted and rebellious; and every group-member has to do the same.'

In view of the above, it is understandable why the numerical growth of the armed forces—which may occur either through the natural growth of the population, or through the expansion of the state, or through the extension of military service —fosters their hierarchization and monocratization. There can be little doubt that it does. No examples can be found of large armies not organized monocratically. Those which are not so organized are invariably small. The undisciplined hosts of medieval Europe were very small. So were the armed hosts of Homeric Greece, of India during the 'Heroic age', and of China during the Chou era. All of them were addicted to a disorderly manner of fighting, their battles consisting of individual duels. Aristotle says explicitly that it was the birth of discipline which enabled large numbers to take part in combat. The evidence drawn from the history of modern Europe is inconclusive because the standing armies of European monarchies were well-disciplined before the introduction of general conscription. But the coincidence of the extension of military service with the assertion of the ruler's control over the armed forces can be observed in a number of cases. There is a strong presumption that the two are structurally connected because all other conditions were widely dissimilar. Mehmet Ali, for example, disbanded the Mamluks, a small corps of hereditary warriors addicted to sedition, and in their place created a large and obedient army of conscripts. This was an essential step in his efforts to transform a *de facto* warriors' republic into a monarchy. The same happened later in Turkey, when

Mahmud II destroyed the Janissaries, who by this time, in contrast to their earlier character, had become a hereditary, restricted and seditious body. Similar was the development which took place in Russia, when Peter I created a large conscript army, and dissolved the corps of Streltzy—a small force of uncontrollable, professional soldiers, mostly hereditary. Again, the large conscript army instituted by the Meiji Reforms in Japan proved a much more obedient instrument in the hands of the government than the caste of samurai, which it replaced. Conversely, the disappearance of large conscript armies and the monopolization of arms by the nobility contributed to the decline of the royal authority during the Heian era. These coincidences found in societies which had little else in common can hardly be accidental, and unmistakably point out to a necessary connection. It must be remembered, however, that the size of the armed forces is only one of the factors determining their hierarchization, and we cannot be surprised, therefore, if we see that in some cases extension of military service may be followed by the spread of disaffection in the armed forces. Such was the case of the Habsburg Monarchy, which in the era of adamantine nationalisms could only have survived if buttressed by an army of professional soldiers.

As remarked above, the growth of the armed forces may be due to the natural growth of the population, or to the expansion of the boundaries of the state, or to the extension of military service, or any combination of these; the shrinkage of the armed forces may, of course, be due to the opposite processes. The existence of these three alternatives naturally complicates the relationship of the size of a political unit to its structure. But long ago the ancient Greek philosophers realized that the two are connected, as can be seen from their discussions of the problem of the optimum size of a city. The connection between the size of the state and the concentration of power did not escape the keen insight of Montesquieu: he states explicitly that republican constitutions are suitable for small states only, whilst great empires must be despotic. Constitutional monarchy is, according to him, suitable for states of medium size. Broadly speaking, this generalization is acceptable if we divest it of its legalistic terminology; as is well known, the appellation has little to do with the distribution of real power. But it seems indisputable that there is a definite and positive connection between the size of political units and the degree of their hierarchization and monocratization. Political structures altogether devoid of any traces of monocracy and hierarchy never comprise great numbers of individuals. Very small primitive bands are never monocratic. All the large states of the past, on the other hand, were ruled monocratically. When the Roman state, for example, grew from a city into an empire, it ceased to be an oligarchy and became a monocracy. True, this regularity is not entirely due to the effects of mere size but also to the consequences of the way in which its increase normally comes about; all large states have been erected through conquest.

There are, naturally, many other factors which determine the degree of monocratization and hierarchization, as can be seen from the example of some large contemporary states which are considerably less monocratic and hierarchic than many much smaller states of the past. Nevertheless, all these states, in spite of the application of the principle of representative government, are far more monocratic and hierarchic than the city states of Greece and Italy. It is significant that the one which is least so—Switzerland—is also the smallest.

THE INFLUENCE OF THE METHOD OF REMUNERATION AND EQUIPMENT

The obedience of the armed forces towards the central government is, as we saw, enhanced if they are equipped and paid by it. If they are not, while being scattered over a considerable area and trying to provide for themselves, a tendency towards territorial dispersion of power will ensue. But if other factors are inimical to territorial scattering of warriors—it, as was the case with the Greek hoplites fighting in phalanx, the best tactics require skilful co-operation by considerable numbers of warriors, and therefore their territorial concentration—then, even if they equip and maintain themselves, no tendency towards territorial dispersion of political power can come into operation. The warriors may emancipate themselves from control by the ruler, but such flattening of the pyramid of subordination will not take place in the territorial dimension.

The best example of such transformation is the evolution of Greek cities. The power of Homeric kings was based to a large extent on their monopoly of external trade; with the wealth thus gained they could maintain and equip numerous retainers. When the passive trade with the Phoenicians was replaced by sea-going expeditions of the Greeks, the control of commerce slipped away from the hands of the kings. Independently gained wealth enabled Greek warriors to equip themselves. And gradually, they emancipated themselves from royal tutelage. But the Greek warriors, unlike their medieval counterparts, did not scatter over the territory they ruled, each taking a small part of it as his domain. The tactics of the day did not permit such development and the smallness of these states made it useless. So, they—that is to say, the warriors in cities where they constituted a definite stratum—ruled collectively, not individually, over their subjects. The warriors of Ramses' Egypt and of other oriental monarchies, on the other hand, who also dominated the population collectively but who were equipped and remunerated by the kings, remained under the control of the latter.

The influence of the manner of equipment and remuneration of the armed forces on their inner structure has been treated with unsurpassed mastery by Max Weber. There is no need, therefore, to present the evidence here, as it can be found in his works. His main conclusion, which my investigations fully support, is that self-equipment is conducive to the relaxation of discipline, i.e. to the flattening of the pyramid of subordination. Such flattening, i.e. the dispersion of power may be territorial, like the feudalization of Charlemagne's empire, or non-territorial, like the establishment of Greek warrior republics.

The practicability of the alternative methods of equipment and remuneration has already been discussed earlier, so no more needs to be said on this point.

THE INTRA-HIERARCHIC OSCILLATION

In order to dominate his subject a ruler must have a body of assistants, enjoying substantial privileges. For reasons already indicated, it is impossible to govern any large group whatsoever without the aid of such a body; and they must feel that they have a common stake with the ruler: that his interest is theirs. This is particularly true if his authority is neither traditional nor willingly accepted by the masses. That is why, as Aristotle noted, dictators rely to such an extent on privileged troops. Moreover, the specifically political authority could emerge and break through the social organization based on kinship only when leaders converted themselves into rulers by gathering armed followers and organizing them along distinct, non-kinship lines. The 'loose men', unbound by tribal customs, but attached to the chief by personal loyalty, enabled him to elevate himself above his fellow tribesmen. This process was already recognized by Ibn Khaldun as a regular accompaniment of the establishment of Islamic monarchies. Particularly suitable for his purpose were foreigners, who, feeling the hostility surrounding them, were likely to be absolutely devoted to their master.

'In Fiji', says R. H. Lowie (*Social Organization*, London 1950, p. 339), 'it happened that individuals fled from their home district to another where the chief exercised only limited authority over his people, but did have the right to assign wasteland for farming. Accordingly, the newcomers would apply to him for an allotment. This, however, at once put them into a different category from the native commoners, who owned their cultivations, making the aliens directly dependent on the chief. But this greatly enhanced the chief's status with reference to his traditional subjects, since now for the first time he could command retainers not allied to the other commoners. . . . In the rise of Kazak and Mongol khans a comparable phenomenon was conspicuous. . . .' To these, innumerable other examples could be added, ranging from Ontong Java and ancient Japan to the Nupe Kingdom in Nigeria and the French Monarchy: the phenomenon is undoubtedly typical. It must be remarked, however, that there was also another reason why many governments employed foreign mercenaries. Before the invention of gunpowder, savage bravery was the most important quality of a good soldier; and it was unlikely to be very common among long-pacified, downtrodden populations. Good soldiers could, as a rule, be found only among rude tribes, inured to violence, outside well-policed areas.

Privileges of the ruler's assistants may be greater if they are the only warriors, and the masses militarily useless; but they cannot disappear altogether even where arms bearing is a universal duty; there must be officers in an army of conscripts.

The policy of a ruler towards his staff presents a dilemma: on the one hand, he must not alienate their support by diminishing their privileges, while on the other hand, these privileges constitute a very real threat to his authority. A despot, therefore, must bestow favours, but must also be 'severe with the mighty', as an exhortation of a Pharaoh to his son puts it. This dilemma gives rise to complicated internal dynamics of hierarchic pyramids, the adequate treatment of which would vastly exceed the framework of the present investigation. I hope to deal with this problem at length in a future work, where the factual evidence supporting the generalizations enumerated below will be examined. The brilliant book of Bertrand de Jouvenel—*Du Pouvoir* (Geneva 1947, 2nd ed.) deals with some aspects of this problem.

Briefly, the regularities which are most pertinent to the present theme are as follows: Monocracy has a levelling effect on social stratification. In the face of an all-powerful despot all others are slaves. Absolute authority, moreover, entails the power to choose the assistants: to promote and degrade. It requires and produces, therefore, very intensive and general interstratic (vertical) mobility. In a democratic and egalitarian society interstratic mobility may be very general, but it can hardly be very intensive, because in such a society the stratification cannot be very steep. The essence of oligarchy, on the other hand, is the hereditary appropriation of positions of influence by their incumbents. The denial to the ruler of the right of choosing his assistants, together with other privileges of the oligarchy, generally constitute the most formidable rampart against despotism. But the magnates are at the same time, usually even more so than the ruler, the oppressors of the masses. For this reason an alliance between a despot and the commonalty against the magnates is a very frequent occurrence. Such an alliance underlay the rise to power of the great number of tyrants in ancient Greece; and also of some modern dictators, or semi-dictators like Peron. But perhaps the most revealing instance of this phenomenon is the 'pact' which Ivan the Terrible concluded with the commonalty of Moscow, whereby, in exchange for the promise of unconditional submission, he undertook to curb the magnates' extortions.

Monocracy means, therefore, as compared with oligarchy, a flattening of stratification; but only up to a certain point, because without a privileged staff the ruler would be compelled to rely solely on the goodwill of his subjects, and be unable to squeeze out of them anything that was against their wishes. There is, therefore, a certain point at which the height of stratification is most favourable to monocracy. But it is unlikely to remain there for very long because, apart from anything else, the personalities of successive rulers are unlikely to be equally strong. The waning of the authority of the ruler is usually accompanied by the accentuation of social inequalities. This is the main reason for the existence of the oscillations between feudalism and bureaucratic absolutism which have been observed in the history of so many states. Oscillation between centralizing despotism and oligarchic dispersion of authority, with concomitant changes in stratification, are particularly visible in the history of the Chinese empire, which was peculiarly mono-hierarchic. It was composed of only three important elements: the ruler, the bureaucracy and the peasants; even priesthood, a well-nigh ubiquitous power, was absent. In poly-hierarchic societies, like those of western Europe from the Middle Ages onwards, the distribution of power within each hierarchic pyramid depends, as will be seen in the next section, on its relations to other pyramids; so the intra-hierarchic oscillations become blurred.

From the point of view of the present enquiry the most important conclusion is that all military circumstances which foster monocracy tend to shape the stratification in a particular way: in a previously oligarchic society they tend to flatten it; in a previously democratic and egalitarian society they tend to heighten it; in both cases they tend to enhance interstratic mobility.

CONSTELLATIONS OF POWER CENTRES

Continuing the theme of the previous section, I must make a few remarks about the influence of possible constellations of power centres within which the armed forces may find themselves; necessarily in the same peremptory fashion, anticipating their corroboration and elaboration in a future publication.

The flattening of the pyramid of subordination, i.e. intrahierarchic dispersion of power, need not be territorial. But obviously, territorial dispersion of power must mean the flattening of the pyramid of subordination; in the extreme form—its disintegration. All the circumstances, therefore, which affect the territorial distribution of power must influence the inner structure of the armed forces too.

Non-territorial dispersion of power may be inter-hierarchic

as well as intra-hierarchic; it is intra-hierarchic if, for instance, the ruler loses the control over his staff; it is inter-hierarchic if, let us say, he becomes less powerful because the church becomes stronger. The inter-hierarchic distribution of power is of great importance to the inner structure of all hierarchies belonging to the same constellation. Thus, for instance, the French Kings of the later Middle Ages, by subjugating the Church, strengthened their position in relation to their vassals and administrative staff; while the German Emperors, having exhausted their resources in an unsuccessful struggle against the Church, ceased to be able to control their subjects, and their empire disintegrated. But, on the other hand, if there are no other power centres besides the bureaucracy, so that the ruler cannot play them against it, he may be unable to control it. This is a typical situation in declining etatistic states, such as the Late Roman Empire or the Late Ptolemaic Kingdom.

As can be seen from the foregoing remarks, the possibilities of interplay of units composing such constellations are manifold. In fact, the subject has too many ramifications and complications to be discussed here. I hope to analyse it in greater detail in a future work, dealing with the general theory of power structures. For the purposes of the present argument it is sufficient to realize that the inner structure of the armed forces is influenced, apart from the circumstances existing within them, by the whole constellation of social forces. Among those, ideological factors are very important, and I must add a few words about them.

The functions of the priest and of the military leader are the earliest sources of authority. Sometimes they are united in one hand—sometimes kept separate; with all kinds of possible transitional variations. The discovery of this fact gave rise to an ardent debate as to whether kingship originated in the function of the priest or in that of the war leader. This debate seems to me pointless. It proceeds on the assumption that kingship is a unitary phenomenon, which must have unique origin. Both assumptions are without foundation. In some cases the supreme ruler may have evolved from the priest, in others from the war leader. Moreover, even if his original function was religious he might divest himself of it, and become mainly a war leader; or vice versa. But what matters most from our present point of view is the effect which the exercise of sacerdotal prerogative by the head of the state has on his authority; undoubtedly it strengthens it enormously. For this reason the existence (or absence) of beliefs investing the ruler with magical and religious powers is of the utmost importance. The relationship between ecclesiastic and political hierarchies, their interlacement or segregation, is of equal portent. It seems, for instance, that the separation of political and sacerdotal powers was one of the forces which created the feudal and corporative society of western Europe, and therefore, one of the chief causes of the distinctiveness of Western civilization. The despotic states, on the other hand, are generally characterized by the amalgamation of these powers.

Some psychologists contend that the pattern of family life, by forming the prevalent type of personality, determines political structure. The authoritarian family, according to them, produces an authoritarian political system; and the same is true about their opposites. It is very probable that there is a connection between the two, although it certainly is not so simple as these psychologists believe, because of the influence of other factors. But in any case, the causation in the contrary direction is equally possible. An example of the latter process is afforded by the recent changes in Soviet Russia, where the authoritarian state is imposing an increasingly authoritarian family pattern.

PRETORIANISM

The inability of the ruler to control his troops may lead to the establishment of a warriors' republic. But often this de-hierarchization of the armed forces fails to crystallize itself into recognized laws or customs, and the sway of the armed forces is exercised through sporadic outbreaks of violence. In order to explain this phenomenon adequately we ought to know more about the origin and functioning of norms regulating behaviour, especially political behaviour, and about their relation to the actual distribution of power; when does the latter crystallize into a definite body of laws and customs? This problem is, of course, far beyond the scope of the present enquiry. All that can be done here is to single out certain conditions which seem to provide an especially propitious breeding-ground for the double-edged phenomenon of the armed forces, insubordinate towards their ruler, dominating society.

The Roman imperial guard stands as a classic example of the

rule of soldiers. But it is by no means an isolated case. Quite the contrary: pretorianism is an exceedingly common phenomenon and can be found in contemporary Syria or Paraguay, as well as in Ancient Egypt or the Arab Caliphate. But what exactly is pretorianism? The term is certainly not applicable to, let us say, a Germanic tribe where warriors elected their chief and decided upon war and peace. The mark of pretorianism is that it is the rule of soldiers exercised not along customarily or legally recognized, constitutional channels, but through mutinies and *coups d'état*. What are the conditions, then, which foster the growth and perpetuation of this phenomenon?

The first observation is that it is always professional soldiers, never conscripts or militiamen that are the driving force of pretorianist insurrections. In China, for instance, the conscripts might have participated in great revolutions, but the various *coups d'état* were perpetrated by mercenaries. The Arabian Caliphate, the land *par excellence* of pretorianism, relied exclusively on mercenaries. Ottoman Janissaries, Russian Stryeltzy, Nubian troops of the Pharaohs, various Latin-American armies—in fact, all armed forces which exhibited pronouncedly the pretorianist features—were professional.

Moreover, pretorianism flourishes best among troops which are not imbued with any particular ideology, but are strictly mercenary. The Janissaries became mutinous only after the ardour of the holy war disappeared from their ranks. The Nazi S.S. and the Soviet police formations (M.V.D.), being thoroughly indoctrinated, remained extremely obedient. All the military formations mentioned before, on the other hand, had a decidedly mercenary character.

The armed forces are more likely to be arbiters of politics if they are the main pillar of authority; that is to say, if the government rests principally on naked force and not on the loyalty of the subjects. Such regimes were particularly prevalent in the Moslem Near East, and significantly, it was the classic land of pretorianism. The armies of the European kings, on the other hand, remained on the whole obedient; one of the reasons being that they were not the sole pillar of the royal authority, which could count on the loyalty of the subjects.

Military might is likely to be a decisive factor in politics in a society where there are no crystallized and universally accepted beliefs about the legitimacy of power: where there are doubts and disagreements about who should occupy the positions of command and what orders he is entitled to give. Thus the rule of primogeniture, unquestioned in European monarchies, gave them the stability which was lacking in the oriental states; and this is another reason why the role of armies in the politics of European monarchies was comparatively subsidiary. The armies of Latin America displayed no pretorianistic tendencies so long as they served the kings. It was only with the establishment of the republics, the constitutions of which did not command universal allegiance, that they became arbiters of politics.

If the masses are politically apathetic professional officers may execute a putsch even with the aid of conscripts. This was shown dramatically in Poland in 1926, during the mutiny through which Pilsudski and his henchmen got into power. He used for this purpose troops recruited from the north-eastern provinces, the composition of which reflected the social structure of those lands. The officers were drawn from the Polish gentry who supported him. The ranks were recruited among illiterate peasants, the majority of whom were devoid of any national consciousness whatsoever. The greater part of the middle class and of the urban proletariat consisted of ghetto Jews absolutely indifferent to the Polish (or any other) state. The troops which were ready to fight for parliamentary institutions and opposed Pilsudski, came from western Poland, where there was a Polish middle class and wage-earners, and where the peasants were literate and conscious politically.

In Spain the pretorianist 'pronunciamentos' became a part of the accepted order, in spite of the conscription through which the army was recruited, owing to the illiteracy and the political apathy of the masses from which the recruits were drawn. When the latter came to have political ideas the 'pronunciamentos' ceased to be easy; and the last of them has kindled a bloody civil war.

It must be noted that a crystallized system of political loyalties is, as a rule, absent in ethnically heterogeneous societies, like those of many countries of Latin America and the Near East.

As we saw in one of the preceding sections, the prominence of the role of armies in politics depends also on their size and importance to national existence.

It also seems that, as Mosca maintains, the socially homogeneous armies are more likely to dominate politics than those split into closed strata. The armies of modern Europe, according to him, played only a secondary role in politics because they were divided into gentlemen-officers, who felt solidarity with the civilian ruling class, and volunteers, recruited among vagabonds, or conscripted peasants. There certainly is a great deal of evidence supporting this thesis: pretorianism became particularly acute in Rome when promotion from the ranks to the highest posts became common; the classically pretorianistic armies of the Islamic Near East were stratically homogeneous, in the sense of not being split into closed antagonistic strata.

Ethnic homogeneity also contributes towards making mercenary armies more dangerous to their employers. The Byzantine emperors recognized it clearly and were always careful to have their mercenaries divided on ethnic lines, so that they could not combine against the government.

V

The Extent of Governmental Regulation

HERBERT SPENCER formulated a generalization according to which militancy, i.e. orientation of a society towards war, inevitably produces the extension of governmental control over the lives of its subjects. He also predicted that such extension would take place in Europe in this century, with socialist ideas serving as its justification, if a series of major wars occurred. The contemporary reader does not need convincing that this prophecy came true. We all know of innumerable laws giving extraordinary powers to governments, which, though promulgated in response to the needs of war, have never been repealed when this emergency was over. We see how not only war but even a threat of war can lead to a curtailment of constitutional freedoms. Nevertheless, although we see in our own days that militancy is accompanied by the extension of governmental regulation, we cannot on the basis of this observation alone say whether the connection between the two is necessary, or whether it is just a coincidence. Nor can we say whether it is of universal nature, or whether it is due to the peculiarities of modern warfare. Another equally important question is: can the extension of governmental control be due to causes other than military?

As to the last question there can be no doubt that the answer must be affirmative. Not only war but also many other kinds of calamities, such as floods, droughts and epidemics, may spur a government to undertake new activities and intervene in many new spheres. The same applies to economic crises. During the depression of the thirties, some governments were forced against their wishes to interfere in the working of the economy. Etatistic policies pursued by the rulers of the Late Roman Empire, or in China by Wang Mang, were not motivated by any explicitly etatistic ideology but by the desire to remedy violent movements of prices and other processes, which were dislocating the economy. Apart from emergencies, the mere increase in the complexity of economic organization makes a wider measure of co-ordination inevitable. In ancient riverine civilizations the necessity of co-ordinating irrigation rendered extensive governmental regulation unavoidable. The complexity of the industrial economy makes it even more imperative. Some other considerations which have similar effects can also be mentioned: such as the need of conserving irreplaceable resources, and the necessity of providing various services which are unprofitable from the business point of view.

A full analysis of the causes of the trend towards etatism, and its extreme form—totalitarianism, is beyond the scope of this study. Nevertheless a few remarks about some other factors, which reinforce this trend in the modern world, must be made.

The completely untrammelled exchange economy seems to be unbearable even to its chief beneficiaries. Competition, however beneficial its effects may be in some ways, means insecurity; and nobody likes to feel insecure. That is why there is a perennial tendency to restrict competition: whether by the 'closed shop' on the part of the wage-earners, or by 'gentlemen's agreements', cartels, etc., on the part of the capitalists, or by guilds on the part of artisans and merchants of all times. But these devices may not suffice without legal backing. Even powerful cartels and trusts feel safest when they have secured legal rights of monopoly. State interference may be called for, then, in order to 'freeze' everybody in his possessions or job. The demand for such 'freezing' is imperious in times of crisis, when the whole machine of commercial economy is flying to pieces. Etatism need not be egalitarian. And the rich are not always against it. On the contrary, sometimes they may positively welcome it as did the monopolistic companies of the mercantilist era and their more recent counterparts. Etatistic interference may, in fact, buttress the privileges of the rich, and be directed against the interests of the poor, as were various measures enacted in Elizabethan England. Numerous laws operating in contemporary South Africa and Southern Rhodesia serve the purpose of protecting the position of the white ruling caste against the free play of the market. The rich are necessarily only against etatistic policies which aim at the levelling of the distribution of wealth—that is to say, socialism.

The desire for equality, at least latent, is ever-present among unprivileged masses, and is quite understandable. What requires explanation is the fact that this desire produces the claim for etatistic intervention; this is undoubtedly due to the peculiarities of the modern industrial economy. Egalitarian movements, often debouching into revolutions, have not been unknown in the past. But their goal was the confiscation of the property of the rich, its distribution to the poor and the abolition of debts. Such a goal was practicable when wealth consisted of hoards of coins, or of lands or workshops which did not constitute productive units, but were united only by being the property of one person. With modern industries such a solution is impracticable: they cannot be broken up and redistributed among the poor. Therefore, the most obvious remedy appears to be their 'statisation', which makes them nominally the common property of all citizens. I cannot enter in this place into the question whether this remedy leads to the expected results or to what results it does lead, beyond saying that although it does no doubt diminish inequalities based on the possession of wealth, it may produce new inequalities based on positions in the bureaucracy.

When we ask why this wish for etatistic levelling is heeded in so many countries, we encounter again the influence, though indirect, of military factors. As indicated in Chapter I, the military developments of the last century gave the masses new power, and therefore their wishes to be protected from the unregulated play of the market and to see the privileges of the property owners curtailed had to be considered. So, these developments produced in a round-about way the extension of governmental control, apart from that directly due to exigencies of war.

A great number of advocates of etatistic policies are recruited from among the aspirants to power and riches, handicapped by non-possession of property. By undermining the importance of the latter and elevating the status of bureaucratic posts they can circumvent this obstacle. Expansion of etatism, therefore, can be viewed as an aspect of the circulation of ruling groups, particularly as the expansion of bureaucracy means multiplication of posts, and therefore increased opportunities of entry for candidates and of promotion for insiders. These two motives have been influential in all clamours for the expansion of bureaucracy, as well as of the army and many other bodies. Bureaucracy, moreover, like any other body, has a tendency to acquire as much power as it can, irrespectively of whether this serves any widely recognized need or not. Whether it succeeds in this endeavour or not depends primarily on the balance of power between itself and various other groups, some of which may be its allies, others its opponents. Constellations of interest and pressure groups may be extremely involved, and it would take us too far to attempt to analyse this problem here.

Much could be said about the psychological aspects of the spread of egalitarian and totalitarian ideologies: about the wish, often unconscious, of isolated human atoms, deprived of roots in any intermediate groups, to be integrated into a mystic body of a party or of a state; about the way in which these new religions displace traditional ones because they are better adapted to pseudo-scientific mentality; about the manner in which the most disparate motives—altruism, lust for power, discontent, messianistic expect-ations, sheer nihilistic destructiveness—are welded together into the united force of a conquering religion. But I cannot pursue this theme any further. For some additional points of view I may refer the reader to Jules Monnerot, *Sociologie du Communisme: Psychologie des Religions Séculières* (Paris 1949); and to my article, 'Are Ideas Social Forces?' (*American Sociological Review*, December 1949). Enough has been said, however, to show that not all extension of governmental control must be due to military factors. We can pro-ceed now to the next question: does warfare or preparations for it always lead to such extension, or is this the result of some particular forms of warfare and military organization?

In view of a formidable array of evidence adduced by Herbert Spencer in his *Principles of Sociology*, and also by P. Sorokin in *Man and Society in Calamity* (New York 1942) and in *Social and Cultural Dynamics*, vol. 3 (New York 1937), there can be no doubt that wars do often produce the extension of state control over the economic and other aspects of social life. Nevertheless, it does not appear that they always do so. Obviously wars between small tribes, with rudimentary political organization, cannot have this effect. But even if we confine our attention to large states, we find that, for instance, in the Ptolemaic kingdom, which was one of the most etatistic states that ever existed, the exhausting wars against the Seleucids and the Romans, by dislocating the bureaucratic machine, forced a retreat from the full control of the economy by the state. The same thing happened in Byzantium in the 12th century. Moreover, some of the most warlike states in the world were not totalitarian, nor even etatistic: the Ottoman Empire, for example, or the Sultanate of Delhi, or the Arabian Caliphate. These states arose through conquest and their whole organization was directed towards war. Nevertheless, they did not regulate the everyday life of their subjects in any marked degree: village communities, castes and guilds conducted their affairs according to age-long custom, and without much interference by the central government, provided the taxes were paid. This non-interference, it must be noted, was not based on any recognition of rights to autonomy, but was the consequence of indifference. The aim of the conquerors was the acquisition of wealth; and if income could be continuously drawn without too much administrative work—then so much the better. Neither the ruler nor his warriors were interested in undertaking such work, except when unavoidable. Nor could socialism—that is to say, etatism based on the ethical condemnation of inequalities of wealth created by the play of the market forces, and aiming at their attenuation—ever develop in conquest states. At most it could exist in the form limited to the ruling stratum, as in Sparta.

The most important cause of this restraint on the part of the states mentioned above was, however, the fact that no military advantage could be gained from subjecting the affairs of all their subjects to strict control by the government, because the mainstay of their military might consisted of professional soldiers; and the only task of the civilian population was to support them. M.P.R. was low in these states not only because they were founded upon conquest. The strength of the armed forces of the Ottoman Empire lay in their mobility, which they owed to great herds of horses and camels. Cavalry constituted the core of the army, in spite of the brilliant role which the foot-soldiers, the Janissaries, played in many battles. The situation in Persia, the Delhi sultanate, and the empire of the Moghuls was similar, though in the last-named the infantry was more numerous. Consequently, the peasant masses were militarily useless, and could without disadvantage be left alone, after surplus wealth had been extracted from them.

The empire of the Incas, on the other hand, was no doubt one of the most etatistic states of the past. It could not be called socialist because its rulers had no intention of promoting equality, but they did regulate even the minutest details of the lives of their subjects. Another attempt in the same direction, though neither so far-reaching nor so enduring, was the reform of the Ch'in King-dom (before it embarked upon the conquest of what is today known as China) attributed to Shang Yang. It is significant that the military organization of both states, like that of their modern totalitarian counterparts, was based on universal conscription; even of women in the case of Ch'in. The fact that the infantry, simply armed, was the mainstay of their armed forces was decisive. The whole populations were militarily utilizable, and therefore had to be strictly controlled; just as in the warring states of the contemporary world they were made into armies.

The treatise on the art of government, presumably written by Shang Yang, throws peculiarly lurid light on contemporary total-itarianism (see *The Book of Lord Shang*, trans. and annotated by J. J. L. Duyvendak, London 1928). The policies which he advo-cates curiously resemble the practices of the Soviet state. But the chief interest of this treatise lies in its unabashed frankness about the motives of their adoption, which are not veiled by the verbosi-ties of propaganda. The policies in question are presented as the best means of enhancing the might of the state, which is con-sidered as the ultimate end in itself. Shang Yang insists on the desirability of the levelling of inequalities of wealth: '. . . when the rich are made poor and the poor rich, then the state will be strong.' All activities should be subordinated to the waging or preparation of war. There should be no way of improving one's position other than by serving the ruler. People should not be allowed to become too opulent lest they cease to become willing to fight. Altogether a faithful picture of modern totalitarian rule.

From the times of Shang Yang onwards, owing to the growing importance of the cavalry, the professional element in the Chinese army was gradually becoming predominant. This process began in the time of Shi Huang-Ti, and went on until the T'ang. As we should expect, the governmental regulation was becoming less extensive at the same time. During the time of Shang Yang the Ch'in kings considered it their duty to provide every one of their subjects with a piece of land. Periodic redistributions of land were also carried out by the Han emperors. Under the T'ang such distributions are becoming rarer, and under the Sung they dis-appear: the emperors cease to protect the peasants against the moneylender. The two significant etatistic and egalitarian attempts which were made in China—those of Wang Mang under the Han dynasty, and of Wang An-Shih under the Sung—were both connected with the desire of their initiators, particularly evident in the case of Wang An-Shih, to reform the army by restoring the peasant militia to its former position as the main pillar. On the other hand, the governmental regulation of the economic life was least extensive under the Mongols and the Manchus, who permitted only very few Chinese to bear arms.

The present regime aims at the full military utilization of manpower, and is more totalitarian than any system which existed since the times of Shang Yang.

What matters most from the point of view of the present enquiry is the fact that in ancient China totalitarianism was embraced because it proved militarily advantageous. The three civilizations: ancient Chinese, Peruvian and modern Western, have little in common, except that in all three of them arose totalitarian states, which in the first two swallowed their less totalitarian neighbours. Lest false predictions be drawn from these cases, I must remark that the victory of totalitarian states is by no means ineluctable: the 'laissez-faire-istic' Roman Republic conquered etatistic Hellenistic monarchies. The rise of totalitarianism in the civilizations under discussion was undoubtedly connected with another trait which they had in common—very high M.P.R., which in the case of the first two was due to the simplicity of armament, and in the case of modern Western countries is the result of the tremendous productivity of their industries.

The example of Sparta shows how M.P.R. circumscribes the extent of state interference in the lives of its subjects: the despised helots had more personal freedom than their warrior masters—the Spartiates. The same example, and (though in a lesser degree) that of early Rome, shows that totalitarianism is not identical with monocracy. Such 'democratic' totalitarianisms, which could be found in other city states of antiquity too, are suited to small political units, or to those which, though they have grown large, still preserve the traditions formed during the time of their smallness: for largeness, in conditions of frequent wars, fosters monocracy. It seems, nevertheless, that monocracy and oligarchy are more propitious for totalitarianism than democracy, because people are more likely to endure various restraints enforced from above than impose them upon themselves. Thus the forms of warfare tending to produce totalitarianism are common to all Europe; but while in western countries their influence has been in some measure counteracted by traditional institutions protecting freedoms of the individual, in Russia traditions of despotism not only gave to these tendencies completely free rein, but even strengthened them.

It does not follow from the foregoing that no state with a low M.P.R. can be totalitarian. It is sufficient to point to the example of Ptolemaic Egypt to show the falsity of such a contention. There, the armed forces consisted of Greek mercenaries, the natives (as is the rule in conquest states) being excluded from military service; and the sole aim of the etatistic policies was to wring as much wealth as possible from the subjects. Moreover, totalitarianism can come to prevail even in a situation in which neither military nor economic nor political eliciting factors are present, but where its sole driving force is fierce religious intolerance; that was the case with Geneva under Calvin. It must also be remembered that well-developed administrative technique is a prerequisite of totalitarianism, particularly in large states.

It may be noted marginally that there seems to be no correlation between the extent of state control in the economic sphere, and in the field of sexual behaviour. The present-day militaristic and totalitarian regimes (and the same was the case with Sparta) permit far greater laxity in sex matters than the 'laissez-faire-istic' states of nineteenth-century Europe.

From the preceding analysis the conclusion emerges that not all tendencies towards the extension of the sphere of governmental regulation are the result of military factors. Nor does intensive warfare in itself necessarily produce totalitarianism. It only inevitably does so when technico-military circumstances make the co-operation of the whole adult population imperative. In conditions of industrial warfare this co-operation is, more than ever, essential. As far as military factors are concerned, the extent of governmental regulation is determined by the incidence of war and the level of M.P.R.

VI

M.P.R. and Ferocity of Warfare

THE complaint is often made that universal conscription did
not promote 'democracy'. This view is based on the vague
use of the term 'democracy'. There is a tendency nowadays
to use the words 'democratic' or 'fascist' where our less slogan-
ridden ancestors would say good or bad. The question whether
conscription made society good or bad cannot be answered on
scientific grounds alone as it involves normative judgments. But
there can be no doubt that it fostered egalitarian reforms. It did
not, of course, promote the 'cause of peace', but that could be
expected only on the ground of the absolutely false preconception
that humanity consists mainly of meek and reasonable beings, who
are enticed or driven into wars by a few villains, be they capitalists
or dictators or kings or whatnot. Innumerable facts plainly contradict such a notion. Some of the most warlike societies, such as
various tribes of head-hunters and cannibals, have no rulers whatsoever: no capitalists or kings who could goad others into battle,
while they themselves sat snugly behind the fighting lines. Democratic Athens was the most aggressive of Greek cities. The people
of the United States were engaged for more than a century in a
continuous war of conquest against the Indians, thus proving
themselves more warlike than any despotic country of Europe or
Asia. The Boer republics, perfect examples of egalitarianism, were
notoriously more aggressive and ruthless in relation to the natives
than the British colonial administration. Hitler rose to power in
consequence of elections fairly conducted. His victories enraptured
ordinary Germans, just as the triumphs of Napoleon delighted the
French and those of Kitchener the British populace. The fund of
cruelty residing in ordinary people reveals itself at such occasions
as lynching of Negroes, 'pogroms' of Jews, religious massacres in

India, or any other riots of this nature. The fact that so many people take pleasure in killing animals is also significant.

It does not follow, of course, that all men are sadists. The distribution of cruelty and compassion, like that of so many other characteristics, probably follows the normal distribution curve; that is to say, there are few complete sadists and equally few perfectly altruistic individuals, the great majority falling between the extremes—neither very gentle nor very cruel but quite ready to use violence if it suits them. These traits are certainly susceptible to modification. It is thus possible to further the cause of Pacifism or bellicosity by inculcating appropriate virtues, but mere democratization, in the proper sense of the widening of the distribution of political rights, is neutral in this respect.

In certain societies at certain times the rulers may be (from the humanitarian point of view) worse than the average (the example of Nazi Germany immediately comes to mind), but in other times and places they may be better. Many generals, for instance, restrained their soldiers from massacring and pillaging conquered populations. It seems, however, that in societies where positions of authority can be attained only through strenuous and ruthless struggle, particularly power-thirsty individuals will be selected for these positions. Nevertheless, generally speaking, the difference between the powerful and the common man lies not so much in moral character as in the consequences thereof. A sadistic dictator may cause the death and suffering of untold millions, whilst a cruel man-in-the-street can only make his wife's and children's lives a misery, and mercilessly beat his dog.

We found no reason to think that the extension of military service in itself blunts or sharpens bellicosity. But it seems, nevertheless, that such extension is conducive to greater ferocity in war. Thus wars waged by tribes in arms are frequently terribly ferocious, often ending with extermination or cannibalistic feasts. On the other hand, where war is the prerogative of nobles, we find it usually regulated by a code of honour. So it was in India during the so-called Heroic Age, in medieval Europe, in Homeric Greece, in China during the feudal period. The change-over to mass-armies, which occurred in China during the so-called times of the Warring Kingdoms, coincided with the great intensification of ferocity. The same happened in Greece when the hoplite replaced the noble charioteer. In early modern Europe the Swiss, whose armies always consisted of commoners, were renowned for their savage manner of fighting. The professional soldiers of the absolute monarchs of eighteenth-century Europe displayed considerable chivalry towards their enemies, best exemplified by the battle of Fontenoy which began with the French and British commanders inviting each other to open fire first. The mass armies of the French revolution put an end to such niceties, and started the crescendo of ferocity, as yet unfinished.

The explanation of these facts seems to be that professional warriors develop some sort of solidarity with their opponents. They come to be conscious of belonging to the same category, as distinct from non-combatants. No such feelings of solidarity across the front can strike root when nations-in-arms are fighting. There, all bounds of the 'consciousness of kind' coincide at the front, and help to inflame hatred.

Needless to say, there are many other factors which influence the ferocity of warfare: such as the degree of cultural similarity or dissimilarity, the importance of the stakes, the prevailing morals; to mention only the most prominent. For further discussion of these last points I may refer the reader to R. S. Steinmetz, *Soziologie des Krieges*.

VII

Classification of Forms of Military Organization

I N order to propound a useful classification of forms of military organization, I am obliged to introduce in this chapter a number of neologisms. I am not a devotee of jargon and I am fully aware of the fact that many pseudo-scientific terms, bandied about by quite a number of sociologists and psychologists, are merely pompous substitutes for perfectly serviceable ordinary words, only more obscure and vague, used in order to dazzle unsuspecting laymen with fictitious knowledge. Nevertheless, no science can progress without developing its terminology, and ordinary words are quite inadequate for designating varieties of social structures. And in building a really scientific terminology we should not pay much attention to linguistic purists and aesthetes. Even if our terms are as ungainly as 'monobromobenzene' or 'trinitrotoluene' . . . it does not matter, provided they help us to order our field of observation. Nor should we be unduly concerned about the increasing abstruseness of our science; for there comes a point in the development of every science when it ceases to be understandable to the uninitiated. It is furthermore definitely preferable to invent new words than to twist the meaning of words already existing, and so produce confusion.

Naturally, a terminology may be good or bad—that is to say, helpful or obstructive—and that depends primarily on its precision, and on whether it conveys something that cannot be easily conveyed with the aid of everyday language. Precision of connotation, it must be noted, however, does not imply precision of denotation in the field of social phenomena, which are fluid and shade into one another. The result is that nearly all concepts describing structural varieties must be what Max Weber calls ideal types: that is to say, extreme, pure types, the logical extremes of possible, continuous variations, which are almost never found in reality without some admixture of traits logically belonging to another ideal type.

The types of military organization singled out in the classification which follows are such ideal types; they are reference-points in relation to which any actual form of military organization can be classified. They have been arrived at by combining three criteria, which—I must stress—have been chosen solely because of their sociological significance; obviously, many other classifications on different bases, say, according to tactics or weapons employed, are possible.

The first characteristic of military organization selected here for the purpose of classification is M.P.R., whose importance has been shown in Chapter I. In principle this variable is amenable to quantification, and, logically can vary from I to O; but in the following analysis we shall deal with two arbitrarily defined extremes—high and low M.P.R. I shall not attempt to determine where exactly they begin and end. Little could be gained from such exact delimitation because in most historical cases the exact value of M.P.R. cannot be ascertained. We are compelled, therefore, to base our reasoning on broad but distinctly perceptible contrasts. The Ch'in kingdom or any of the European participants in the last war provide examples of undoubtedly very high M.P.R. On the other side, the M.P.R. in pre-revolutionary France or the Ottoman Empire must be classified as decidedly low.

The second criterion selected here is the degree of subordination. Armed forces may be strictly subordinated to their leader, like any of the contemporary European armies, or they may consist of a motley of practically independent warriors. True, subordination can hardly ever be altogether absent; all bodies of warriors have chiefs, even if elected only for the duration of a campaign. Nevertheless, the extreme of the complete lack of subordination is sometimes very nearly attained; the armies of the Crusaders provide an example. Moreover, if we consider the usual manner of living of warriors, and not only their behaviour during a campaign, then we can find many cases of almost complete lack of subordination: warriors' republics, where every warrior was equal to any other and no one possessed authority, existed, for example, among the Cossacks of the Dnieper, among the Masai of Africa, and (in a less extreme form) in Sparta.

The degree of cohesion is the third criterion. A military body may be closely knit, tightly organized, or it may constitute an amorphous multitude of warriors, independent of each other, hardly maintaining any contact. Any modern army or the Spartans can be taken as an instance of the former possibility; the knights of medieval Germany or Poland exemplify the latter. It is true, of course, that a group without any cohesion whatsoever ceases to be a group. So the logical extreme of a completely uncohesive body of warriors cannot materialize. Nevertheless, it can be very nearly approached.

Cohesion, it must be noted, is not identical with physical, geographical concentration, though the latter favours the former. Geographical dispersion can coincide with high degree of cohesion, as can be seen from the example of the British army, which remains a very cohesive body in spite of having outposts in Hongkong and Singapore. Physical closeness, on the other hand, does not in itself produce organizational cohesion, as can be seen from the example of the Tallensi tribe in Africa who, though living in close proximity, constitute a very uncohesive body.

Subordination implies cohesion but not vice-versa. Obviously, strict subordination means tight organization and, therefore, cohesion. But a group may be tightly organized on egalitarian, non-hierarchic lines. The Spartan homoioi are an example of such a body. It is true, nevertheless, that in geographically dispersed bodies cohesion can generally be assured only through subordination, and any weakening of the latter produces a weakening of the former too, as happened in the case of the disintegration of the Carolingian Empire. But social consequences of geographic distance depend on facilities of transport and communication. So, to be exact, we should speak of culturo-geographical distance or dispersion. Perhaps the term 'Pheric' (from the Greek word— to transport) could be used instead. Pheric distance could be defined as distance measured by the time in which it can be covered.

Let us designate high M.P.R. by M, low by m; high degree of cohesion by C, low by c; high degree of subordination by S, low by s. By combining the three criteria we obtain the following types of military organization:

(1) msc; that is to say characterized by low military participation ratio, low degree of subordination and low degree of cohesion. The examples most nearly approaching the logical extreme are: the parts of medieval Germany dominated by Raubritters, the Polish kingdom during the late Middle Ages, the Radjput states in the thirteenth century. I propose to call this type of military organization 'ritterian'.

(2) MsC; that is to say, high M.P.R., low degree of subordination and high degree of cohesion. The examples most nearly approaching the logical extreme are: the Cossack settlements in the days of their independence, the Masai tribes in East Africa. I propose to call this type 'masaic'.

(3) MSc; i.e. high M.P.R., high degree of subordination and low degree of cohesion. No form of military organization possessing these traits can ever come into being because, as subordination implies cohesion, a combination of high subordination with low cohesion is impossible.

(4) Msc; i.e. high M.P.R., low subordination and low cohesion. Examples: the Tallensi tribes, the Trekboers in South Africa, the North American frontiersmen. I propose to call this type 'tallenic'.

(5) MSC; i.e. high M.P.R., high subordination, high cohesion; e.g. European participants in the two World Wars, the Ch'in kingdom in China, the Old Kingdom of Ancient Egypt, I propose to call this type 'neferic'.

(6) msC; i.e. low M.P.R., low subordination and high cohesion; e.g. Sparta and some other Dorian conquest states. I propose to call this type 'homoic'.

(7) mSC; i.e. low M.P.R., high subordination and high cohesion; e.g. Egypt under the early Ramsesides, the Abbasid Caliphate in its early days, Prussia before Stein, or any other absolute monarchy, supported by disciplined professional troops. I propose to call this type 'mortazic'.

(8) mSc; i.e. low M.P.R., high subordination and low cohesion. In view of the incompatibility of high subordination with low cohesion, this combination is impossible.

We get thus six pure types of military organization: neferic, mortazic, homoic, ritterian, masaic and tallenic. I must repeat that all real forms of military organization are a mixture of these pure forms, some not approaching particularly any of them. Thus, for instance, the Byzantine military organization was predominantly mortazic under Justinian I, it acquired neferic traits under the Heraclean dynasty, and towards the end approached the ritterian type. The military organization of the Chinese Empire oscillated between mortazic and neferic types, that of the Ottoman Empire between mortazic, homoic and ritterian.

The three main determinants of military organization are

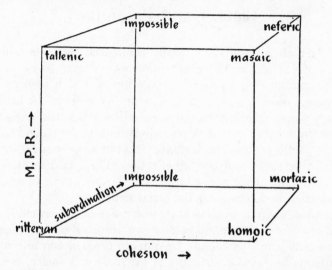

represented by the geometrical dimensions as depicted by the edges of the cube. Their numerical values go from o to 1. The spaces in the proximity of six of the points of the cube represent our six extreme types of military organization. Most of the real cases could be represented by clusters of points nearer the centre. Transitions from one type to another must by definition be caused by factors acting upon these determinants.

VIII

Biataxy and Polemity

NOT all societies are equally influenced by their military organization. There are some which seem to be almost uninfluenced by it—the evolution of which seems to be determined only in a very small measure by shifts in the locus of military power. The substantial attenuation of political and economic inequalities which took place in Britain in the second half of the nineteenth century, for instance, was not accompanied by any marked changes in military organization. The question therefore arises: what are the factors which circumscribe the role of military organization in determining the social structure?

The most obvious of them is the mere size of the armed forces. At the limit of their expansion they can absorb the adult population, and their inner structure becomes in times of war almost co-extensive with the total structure of the society. A society is more likely to be exempt from the influence of the organization of its armed forces if these form a distinct body, separated from the rest of the population; that is to say, if they are composed of professional warriors. But even then, military organization may remain the chief determinant of social structure because these warriors may dominate the whole society. Naturally, other things being equal, the more proportionately numerous they are, the more easily can they impose their will on the rest of the population. The distribution of power within a society may be independant of the military organization only if the armed forces are so small that, given the state of military technique, they are unable to keep down the civilian population. The only exception to this rule is a hierocratic society, so pervaded by religiosity that fear of the priestly power makes even the warriors docile. But armed forces so small as to be unable to keep down the civilian population are unlikely to be sufficient for defence against the external foe. And although we find some primitive tribes which are so isolated that they do

124

not need to worry much about defence, no large state could ever feel secure from external attack for any length of time. No really independent state has ever survived for long without strong armed forces. The states situated on islands, however, could neglect their land forces and rely almost solely on the navy. But the navy, and this fact is full of far-reaching consequences, cannot interfere in the internal affairs of a country in the same way as an army can. Ships are unsuited for street fighting or for repressing uprisings in the provinces. Moreover, because of the heavy outlay of capital per sailor, the numerical strength of a navy must be low as compared with that of an army which the same country could put into the field. For these reasons politics in the states, which could rely almost completely on the navy for their defence, could be singularly free from the interference by military men. The most outstanding examples of such a situation are Britain during the last two hundred and fifty years and Rhodes during the Hellenistic era. Venice and Holland, during the era of their real (as distinguished from mere tolerated) independence, also belonged to this extremely restricted category. Venice was protected from the mainland by lagoons and the Dutch cities by low-lying fields, which could easily be flooded. According to Montesquieu, who lived there, Dutch mercenary soldiers obeyed the city patricians because they feared being drowned if the latter opened the dykes; they were not allowed to enter the cities. Venice, it must be added, was the only Italian city-state which did not come under the rule of the condottieri.

These maritime nations were far from being pacific, but the wars they waged did not undermine their political organization: the armies were far away and could not dominate the metropolis. On the contrary, these colonial wars provided an outlet, innocuous as far as internal politics were concerned, for the excess of adventurous energies which otherwise might be employed in subversive activities.

Security from invasion by land is not in itself sufficient to minimize the role of the armed forces in determining social structure. This is seen clearly from the case of the Japanese society which, in spite of its almost complete immunity from external danger, never ceased to be dominated by the warriors. The part played in politics by the armed forces can be negligible only if violence is not constantly resorted to. Such restraint can prevail only if certain conditions obtain. The first is that a society must be fairly homogeneous; sufficiently homogeneous for the beliefs regulating exercise and transmission of authority to be unequivocal and universally accepted. The customs and ideals of all sections of the population must have a common denominator which permits some *modus vivendi*, without one section coercing another. The fate of the liberated, ex-colonial, plural societies, like Burma or Palestine, or of Spain, well illustrates the point.[1]

Another condition of minimization of the role of the armed forces is economic prosperity. Empty stomachs are generally the source of the most serious outbreaks of violence. This, obviously, is not a new discovery and has been known since times immemorial. Hong Mai, a Chinese writer of the twelfth century, thus formulates a general theory of revolutions: 'Since ancient times appearance of banditry depended on famines produced by droughts or floods. Driven by cold and hunger men gather to rob. If there are at the same times sorcerers who, profiting by the occasion instigate the people, the harm which may result is incalculable.' Naturally, the struggle for power must be more bitter if defeat means starvation or violent death, than if its only consequence is the loss of additional comforts and prestige. That is why representative government, as it is known in north-western Europe and North America, the basis of which is the readiness to compromise and the renunciation of the use of violence, cannot function in countries where millions are starving. And starvation has been a regular occurrence throughout human history; demographic pressure made it inevitable. This is why societies whose basic structure is not determined by military organization are so exceedingly rare. This is also the explanation why Japan, secure from external danger, remained under the sway of warriors. The demographic pressure in that country was very high; the population remained stationary only because of the prevalence of abortion and infanticide. No wonder then, that the struggle for power and literally for existence was very bitter. Particularly grave was the numerical increase of the ruling strata which pushed down their standards of living. The only possible outcome of such a situation was ruthless domination of one section of the population over the rest. Paradoxically, the lack of external wars and conquests seems

[1] This group of circumstances constitutes one of the most important determinants of social structure; I shall refer to it by the name of concord.

to have aggravated the internal oppression. Incidentally, I must draw the reader's attention to the fact that the low birth rate of the ruling classes of western European countries in recent times, whatever its eugenically baneful consequences, undoubtedly helped to soothe the internal struggles. Everybody could attain the position of his father or even higher; nobody had to be eliminated from the charmed circle, as happens in a particularly great measure in countries where polygyny is practised.

Demographic pressure usually leads either to intensification of internal struggles or to attempts at conquest, or both. In either case, the role of the armed forces will grow in importance—except in the case of maritime states, which, because of the peculiarities mentioned above, can embark upon conquests without becoming militarized. Religion is the only power which can subdue, even if never completely, the explosive forces produced by over-population and poverty; so that the issue of 'who gets what' does not need to be settled by naked force alone. Traditional India is the most striking case of a society full of tensions, whose structure is determined nevertheless more by religion than by naked force. It is extremely probable that the weakening of the hold of traditional religion, taking place now, will lead to internal upheavals and/or external aggression: particularly if, because of the improvement of medical services, the demographic pressure becomes even higher, as it will unless the birth-rate drops substantially, thus enabling the accretion of wealth, due to technical improvements, to overtake the growth of the population.

A society in which the distribution of power and, consequently, of wealth, prestige and other desirable things, of ophelimities as I proposed to call them, is settled by naked force, that is to say, by the use or the threat of violence, may be called biataxic (Greek '*bia*'—violence). Naturally, this as well as its opposite—a society in which naked force plays no part whatsoever, and which may be termed abiataxic—is a logical extreme, an ideal type, to which any real society can only approximate but which can never be attained.

There are many other factors besides those mentioned above, which participate in determining the strength of tensions within a society. Thus, for instance, the methods of upbringing, if harsh and unsparing in the use of the rod, together with frustrations generated by norms regulating the family and sex life, may help to augment the fund of aggressiveness, cruelty and hatred. The mere weight of traditions of cruelty and harshness, and the inertia of revenge, must be taken into account too. But the high level of tension resulting from poverty and these circumstances is not the only factor which makes the struggle for power bitter, thus enhancing the importance of military organization. As I said, government not based on naked force can function only if certain beliefs are accepted by the overwhelming majority of the population; if there is an agreement on the right to command and duty to obey. If such agreement does not exist, either because of ethnic heterogeneity or in consequence of an internal schism, naked force must remain the argument of the last resort, and the distribution of military might the principal determinant of social structure.

Herbert Spencer classified societies into militant and industrial. Put in this way this classification appears nonsensical, as there is no shred of evidence that an industrial society (taking the word 'industrial' at its current meaning) must be peaceful. Needless to stress, such a society is by no means immune to bellicosity, and Spencer knew it; he was quite alive to the realities of colonial expansion. His intention, obscured by the infelicitous choice of words, was to bring to light profound differences between societies whose life is oriented towards war, and those preoccupied mainly with peaceful production. There can be no doubt that this factor of the degree of orientation towards war is of radical importance in determining the structure of societies. But before we proceed to examine its influence we must stop to consider the terminology.

The term 'industrial' is obviously altogether misleading. It would be better to designate the opposite of 'militant' simply as 'non-militant'. But even this usage would not satisfy the needs of the analysis which follows, because the accepted meaning of the word 'militant' carries a connotation of aggressiveness; and a society may be oriented towards war defensively. It would be equally unsatisfactory to use the term 'militarized' because it has acquired a meaning different from what I wish to express here: it reminds one of uniforms, the goose-step and all the other paraphernalia of a modern army; we could hardly say without twisting the accepted meaning of the words that, for instance, Navajo society was militarized, even though it was strongly oriented towards war. Moreover, the patterns of army organization may be adopted for purposes other than the prosecution of war. Thus we can say that the creation of the Salvation Army constituted an act of militarization of a religious body. I am obliged therefore to introduce a new term—'polemity' which may be defined as the ratio of the energy employed (directly or indirectly) in warfare or preparations for it, to the total amount of energy available to a society. Complete polemity and its opposite,—apolemity, are, of course, pure polar concepts, seldom attained in reality. Extreme polemity implies the absorption of the totality of the social structure into the military structure; they become identical, as they practically were, for instance, in the Ghazi emirate of Menteshe, from which germinated the Ottoman Empire, or in the society of Suavians, or in various societies of pirates, like those which existed on the Lipari Islands and the Cilician coast in antiquity and on the Antilles and Madagascar in the seventeenth and eighteenth centuries, or of robbers, like the Balkan Hajduks. Sweden and Holland in the nineteen-twenties provide on the other side examples of extreme apolemity. The degree of polemity is not determined by the type of military organization. Thus, for instance, both England and Prussia possessed in the eighteenth century mortazic military organization, but while England was only slightly polemitic, Prussia was the epitome of polemity in Europe of that era. Nevertheless, with the types of military organization characterized by high M.P.R. a higher degree of polemity is on the whole possible than with the types distinguished by low M.P.R., because only in the former case is the complete overlapping of the military structure and the total social structure possible. Furthermore, societies possessing military organization characterized by low cohesion can be only moderately polemitic, because polemity depends on the intensity of warfare, and the latter produces internal cohesion. It follows that the ritterian and tallenic types of military organization can be found only in moderately polemitic societies. In large societies successful prosecution of war demands monocratic co-ordination; only those among them, therefore, which possess neferic or mortazic military organization can be highly polemitic; and, as the possibility of completeness of polemity depends on M.P.R., the highest polemity of a large society can be attained under neferic military organization.

A high degree of polemity means that social structure is militarily determined, that is to say, determined by the form of military organization. The low degree of polemity does not, however, assure an equally low degree of military determination of social structure because of another factor—biataxy, which has been defined earlier as the degree to which distribution of ophelimities in a society is determined by the use or the threat of violence. That polemity and biataxy do not always go together can be seen from the comparison of the Aztec confederation with Tokugawa Japan. The former was highly polemitic but abiataxic; it was oriented towards war but its stratification was based mainly on religion. Montezuma had no police to enforce his will; he, as well as his priests and nobles, were willingly revered by the people. Tokugawa Japan, on the other hand, was an unaggressive state, not very well prepared for defence, whose rulers desired seclusion from the outer world; but they ruled their subjects with an iron hand. It was a state where the seething hatred of the nobles against the shogun, and of the peasants against the nobles, was prevented from debouching into a revolution mainly by terror. There are some reasons for believing that biataxy and polemity tend to be inversely related. This would not be surprising ·because demographic pressure unrelieved by war must sharpen inner conflicts; and psychologically, the fight against outsiders permits a projection of internally generated hatreds on them, thus soothing inner conflicts. Nevertheless, there are societies which are highly biataxic and highly polemitic at the same time: for example the Inca, Ottoman, Soviet and Ch'in Empires.

A society can be free from being militarily determined only insofar as it is neither polemitic nor biataxic. Some societies are almost free from such determination—others are not. Needless to stress, these are again polar concepts, between which lies a whole gamut of intermediate possibilities, differing in the degree of military determination. Neither the former nor the latter constitute a unitary class: the latter differ in their structures as much as do their military organizations; the former agree only in not being militarily determined and a priori can differ in all imaginable ways (they may, for instance, be plutocratic like Victorian England, or hierocratic like the Tibet of the Dalai Lamas). Plainly, even a cursory investigation of varieties of abiataxic and apolemitic societies and their determinants would greatly exceed the bounds of the present enquiry. But I must mention that certain kinds of societies require the freedom from being determined by the military organization as the condition of their existence.

A prominent nineteenth-century English historian Seeley

propounded a 'law' that the amount of freedom in a state is inversely proportional to the pressure on it from outside. The snag in this statement is the uncertain meaning of the word freedom. Societies differ in what they allow and forbid, and it is usually very difficult to estimate where the sum of freedom is greater. If in one society sexual behaviour is almost unrestricted but the criticism of rulers strongly repressed, whilst in another the opposite state of affairs prevails, then it is quite arbitrary to say that the one is a free, and the other an unfree society. Naturally people always accept restrictions to which they are accustomed, whilst restrictions to which members of other societies are subject appear to them outrageous. Ilya Ehrenburg found that the Americans were less free than the Russians because in America if a man stayed in an hotel with a women who was not his wife, the police would break into the room and drag them out of the bed, whilst in Russia they would do nothing about it. One could even argue that the sum of freedom was greater in Hitler's Germany than in Victorian England; after all, Victorians were, according to present standards, 'slaves of conventions', women were kept in tutelage and wage-earners were tyrannized by their employers. It could also be said that political freedom in England is only possible because of the Englishman's rigid adherence to convention, which always strikes continentals; and that a Russian under the Tzars had a much greater choice of permissible kinds of behaviour than an Englishman living under parliamentary government with adult franchise. 'You don't need the police,' said a Russian to Sir John Maynard, 'because you all have mental strait-jackets.' Moreover, freedom for one often means unfreedom for another: if an owner is free to close his factory which is a mainstay of the community, then his employees have no freedom to make arrangements for permanent residence there. Finally, freedom may be invidiously accorded to different sections of the population: in disciplinarian Sparta women were much freer than in 'freedom-loving' Athens. From the foregoing the conclusion inevitably emerges that on strictly scientific grounds we can decide that there is more freedom in society A than in society B only if everything that is prohibited in A is also prohibited in B, whilst certain things which are prohibited in B are allowed in A. Such a situation cannot, of course, be encountered often. In all other cases we can judge relative amounts of freedom only on ethical grounds, on the basis of a conviction that certain freedoms are good and indispensable, whilst others are inessential or even reprehensible.

These considerations do not invalidate the essential point of Seeley's 'law'. Taking the restrictions of his environment for granted, he meant by 'freedom' freedom from control by the government and from arbitrary rule. After such reinterpretation his law becomes almost identical with the generalizations expounded on the preceding pages: namely, that intensive warfare fosters the extension of governmental regulation and monocratization.

The law can be also taken to mean that representative government thrives only in countries which are not under hard pressure from outside. Broadly, this is true as far as large societies are concerned; and the history of the modern occident certainly lends support to this view. Representative government flourished uninterruptedly only in the island of Britain, and later in its former colonies, all of which were sheltered by the oceans and the British navy. The only other country which held to equally radical parliamentarianism—Poland—was destroyed. Evidently, only a monocratic, centralized, militarist state could survive in the centre of Europe. The 'rule of law'—that is to say, a situation where obedience is due to general rules and the sphere of arbitrary command is small, at any rate as far as legal sanctions are concerned—implies some limitations on authority and is incompatible with extreme monocratization, but positively connected with a representative mode of government. If then, we take freedom as meaning the 'rule of law', representative government and restricted governmental regulation Seeley's 'law' is quite correct.

The analysis of the whole problem is gravely impaired by the lack of a general conceptual framework which could serve as a basis for an exact and unambiguous description of social structures. This defect could be remedied only in a general treatise on social structures. So, I must confine myself to a few, necessarily inadequate remarks.

When Spencer spoke about industrial societies he meant more than just the opposite of a militant, or polemitic society. Obviously, he had in mind a type of society similar to the England of his day—commercial, laissez-faire, liberal, contractual, spontaneously mobile. (The qualification 'spontaneously' is very important because compulsory mobility is a very common feature of despotic states: witness the enormous shifts of population carried out by the orders of Shih Huang-ti, Assurbanipal, Inca Patchakutek, Hitler, Stalin and others.) Such a society must be abiataxic and apolemitic; otherwise it could not exist. Without the long peace the nineteenth-century liberalism could not have thriven. True, this was a period of colonial expansion, but colonial wars of conquest, whatever may have been the havoc wrought upon the victims, were waged by the European states in an off-hand fashion; they never demanded total mobilization of resources. And significantly, the main bastion of liberalism was the island of Britain, secure from invasion. The United States, which Ferrero called Europe's enlarging mirror, displayed the features of an open, mobile, liberal society to an even greater degree; and they too were free from the necessity of serious military effort.

In order not to be misunderstood, I must emphasize that I do not maintain that the type of society which, because of the lack of more adequate terminology, I called open, mobile and liberal, must arise wherever there is no need for tightly knit military organization: the absence of such a need is only one of the conditions of its existence. Another indispensable condition is a low degree of biataxy. But even abiatxy and apolemity combined cannot assure the existence of an open, liberal, contractual society. Much depends on the economic situation. In a stagnating or declining economy there is a tendency on the part of incumbents of all advantageous positions to appropriate them, and to form closed groups from which all outsiders are debarred. High social mobility of a spontaneous kind is possible only if groups are willing to admit newcomers; and that depends largely on whether the extra mouth is not universally regarded as being much more important than the extra pair of hands. This statement obviously needs further elaboration which cannot be undertaken here.

IX

Interstratic Mobility

PITIRIM SOROKIN in his fundamental work on *Social Mobility* expressed the view that war fosters vertical (or as I prefer to call it—interstratic) mobility; and I believe that on the whole he is right. Nevertheless, much depends on the kind of war. Thus a war between unstratified tribes, which does not lead to subjugation of any of them, can have no such effect. And this type of war is very frequent among pre-literate peoples: it is the rule among hunters and food-collectors, where war is usually enveloped in ceremonialism and magic, and the possibility of conquest not even conceived. On the other hand, all wars ending in conquest produce wholesale interstratic shifts. If it is a primary conquest of one unstratified tribe by another, then it means that a pyramid of stratification is erected, and that the conquerors turn from ordinary men into masters (e.g. the establishment of several Hamitic kingdoms in East Africa). If the conquered society was already stratified, then a conquest may result in the superposition of an additional stratum (e.g. the T'opa conquest of North China or the British conquest of India); or in the wholesale degradation, or even extermination, of the old ruling stratum, and its replacement by the conquerors (e.g. the Norman conquest of England, the Nazi conquest of Poland, the Turkish conquest of the Balkans). A conquest may, of course, lead to a more or less complete ejection or extermination of the defeated population (e.g. the Anglo-Saxon conquest of Britain, the European conquest of North America, the Slavic conquest of Illyria). If the last-named possibility materializes, then we can hardly speak of interstratic mobility unless we assign some social status to the dead; but ejection usually results in the degradation of those who occupied elevated positions, and cannot find equivalents in the place of refuge. Also, war must produce interstratic mobility insofar as it is connected with slave hunting; and it can be so connected even in fairly simple societies, not embarking upon conquests, like the Kwakiutl of the north-western coast of America or the Bagobo of the Philippines. Sub-jugation may take many forms, differing in their thoroughness and accordingly in their efficacity in generating interstratic mobility. The defeated may be all turned into slaves or serfs (e.g. the Portuguese colonization of Brazil, or the Dorian conquest of the Peloponnese). In contrast to this stood modern Europe until recent times, where all subjugations were of a very mild kind: all that happened was that people had to pay taxes to a different govern-ment, and only the personnel of administration, sometimes only the top personnel, was supplanted; other people were left in their jobs, possessions, etc. It follows that the effect of these subjugations on interstratic mobility could only be slight; particularly, as the class of officials was not numerous in those days. Sometimes subjugation meant nothing more than the acknowledgment of suzerainty and payment of tribute, as in the case of the Persian conquest of the Phoenician and Greek cities. Sometimes the con-querors exterminate or degrade the recalcitrant members of the old ruling stratum, but allow others to retain their positions, as did the Romans after conquering Gaul, or the Mohammedans in Visigothic Spain. Sometimes the pre-existing ruling stratum is degraded or exterminated, and replaced with the new, drawn from the lower strata of the subjugated population, which fits better into subordinate positions in the conquerors' administrative machine (e.g. the Soviet conquest of eastern Europe, and, it seems, the Inca conquests of certain Peruvian principalities).

Another way in which war may foster interstratic mobility is by dislocating social structure. A hard war may change radically the relative values of different kinds of property, thus enriching some and impoverishing others; it may alter the values of skills, nullify savings, destroy goodwill, close some markets and open others. All such perturbations must cause shifts of individuals on the social ladder. Moreover, interstratic mobility is low if groups, of which a society is composed, are closed. Closure means order and, there-fore, any violent concussions sustained by the social order produce cracks, through which ambitious climbers can infiltrate. The after-effects of World War I in Germany, and of World War II in France provide an illustration of this kind of sequence.

War undoubtedly promotes interstratic mobility within armed forces. In peacetime there is no real criterion of a soldier's ability.

Promotion, therefore, is inclined to be by seniority, tempered by sycophancy and nepotism. Moreover, whilst in time of peace nobody worries whether any military talents are wasted or not, in time of war, particularly a strenuous one, it becomes imperative to find the ablest. This is the reason why usually, when an army enters a war after a long period of peace, a wholesale sifting of commanders takes place. The sifting conducted by Stalin during the last war was truly prodigious, but it was not unprecedented; a similar sifting occurred in the Prussian army after Jena, to mention one of many other examples. Much depends, of course, on the seriousness of the war, the imminence of defeat and the direness of its consequences. For parlour wars, such as those conducted by the European monarchies in the eighteenth century, or those indulged in at various periods in China, aged princes, indolent courtiers and other venerable dignitaries were good enough as commanders. So, generally speaking, we can say that war produces interstratic mobility in the measure of its severity.

Another factor which can promote interstratic mobility within an army is the differential mortality of various ranks; higher mortality among higher ranks means promotion for others. Its existence depends on tactics, armament, the code of fighting, etc. Often the mortality rises as we go from privates to junior officers, and afterwards decreases, being very low among generals, who are more frequently dismissed. If an army constitutes only a small part of the population, and the replacements are incorporated along stratic lines—i.e. if killed privates are replaced by other peasants, officers by other noblemen, generals by other princes (as was done in the armies of pre-Napoleonic Europe)—then the differential mortality may have no influence on interstratic mobility. Such a mode of replacement is impossible if M.P.R. is high. If replacements are always drawn into the lowest ranks, then casualties, whether they are differential or not, must speed up promotions (e.g. all armies participating in World War II).

In societies where warriors constitute the upper stratum, ascension will be fostered only if the rate of casualties is higher than the reproduction rate of that stratum. This may happen (e.g. the wars of the Roses in England), but also may not (e.g. France and Germany in the twelfth and thirteenth centuries). The reader may ask how I know that my statements concerning these examples are correct in view of the lack of statistics. My answer is that if one constantly hears of noble scions roaming over the country in search of fiefs, then, it is obvious that there must have been a surplus of them. It was mainly from this surplus that crusaders were recruited. If heirless fiefs are frequently mentioned, then the opposite situation must have obtained.

As already hinted at previously, monocracy is correlated with high interstratic mobility. The main pillar of a despot's power is his ability to promote and degrade; punishment as a rule presupposing degradation. Once he ceases to be able to do so, once his assistants become a self-perpetuating body, be it co-optative or hereditary, his power is on the wane. All rulers who attempted to overcome an oligarchy did so with the aid of new men—men drawn from the lower strata, who saw their advancement in the extension of the ruler's power. The Bourbons in France, the Romanovs in Russia, the Abbaside Caliphs, the Mughuls in India, the Kings of Dahomey, to cite only a few of the available examples, attained absolute authority by following this policy, and maintained their authority only in so far as they could stick to it. This applies not only to states but to any kind of hierarchy: everywhere the power of the boss is based on his ability to promote and degrade. Admittedly, interstratic mobility depends on other factors too, above all demographic factors; but we can say that, other things being equal, the degree of subordination does tend to co-vary with interstratic mobility. It follows that armies characterized by a high degree of subordination will be interstratically mobile.

It may be said that democracy should also foster interstratic mobility, because it implies the ability of the masses to choose their leaders. This is true, but democracy tends to level social inequalities: a society can be stratified only in so far as it is imperfectly democratic; and where there are no strata there can be no interstratic mobility. Oligarchy, whose permanant tendency is to close itself and prevent entries and expulsions, is the regime which combines high stratification with low interstratic mobility.

In drawing conclusions from the foregoing, we must remember that interstratic mobility may flow along other than military channels, and that societies differ in the degree of their polemity. In a society like the Ottoman, where the state was really an army on the march, and the ruling stratum perfectly polemitic, all interstratic mobility necessarily flowed along military channels. In England, on the contrary, from the fall of Cromwell's regime onwards, the army constituted no elevator whatsoever. Even in Hohen-zollern Germany its role in this respect was minor in comparison with business enterprise. The way in which any change in military organization may affect total interstratic mobility depends on what happens in other sectors of social life: an intensification of inter-stratic mobility within the armed forces may be accompanied by a drop in the total interstratic mobility because of the shutting of other ascensional channels. In the Roman Empire, for instance, the interstratic mobility along commercial channels greatly dimin-ished after the third century, while that proceeding through the army increased very markedly; whether the total volume in-creased or diminished could be established only by a large-scale historical investigation (cf. my article, 'Vertical Mobility and Technical Progress', *Social Forces*, October 1950). Moreover, the results of a process of polemitization depend on the pre-existing situation in sectors hitherto independent of military authorities. Bearing this in mind we can now attempt to examine the influence of particular forms of military organization.

The tallenic and masaic types can hardly foster interstratic mobility, as they provide no ground for the development of strati-fication; stratification may, however, develop in a society whose military organization is of that kind, on other grounds, provided that the society is fairly abiataxic and apolemitic.

The ritterian and homoic types, though having stratificatory influence because of the low M.P.R., do not promote interstratic mobility because of the low degree of subordination; they rather foster a closed oligarchy.

The mortazic type tends to produce steep stratification, in virtue of its low M.P.R., whilst its high degree of subordination fosters interstratic mobility, which may reach quite extraordinary pro-portions, as it did in the Ottoman Empire (see my article men-tioned above). But if the authority of the ruler diminishes and military organization becomes semi-homoic (i.e. pretorianism develops), then the interstratic mobility tends to decline: all pre-torianist troops tend to become hereditary (e.g. the Stryeltzy, the Janissaries, even the Mamluks . . . finally).

The neferic type tends to produce a stratified hierarchy be-cause of its high subordination, which, on the other hand, pro-motes interstratic mobility. This mobility is further fostered by levelling tendencies generated by the high M.P.R.

X

Types of Military Organization and Types of Social Structure

PRELIMINARY REMARKS

BEFORE we examine more closely the factual material bearing upon the problem of correspondence of the types of social structure and the types of military organization distinguished on the preceding pages, let us glance over the effects produced by the three variables in terms of which we have classified the forms of military organization.

M.P.R. affects primarily the steepness of stratification, although its effects are modified by the facility of suppression. Other things being equal, the greater the facility of suppression, the steeper the stratification. Secondly, M.P.R. influences the extent of governmental control. Thirdly, it affects the pervasiveness of the influence of the other two variables.

The cohesion of the armed forces tends to produce the cohesion of the body politic.

Subordination within the armed forces tends to produce subordination in the whole body politic. It has stratificatory effect, as it requires a hierarchy, but on the other hand, it has a levelling tendency. It promotes interstratic mobility.

And now, having restated these principles, we can look closer into the factual material. But at the risk of being tedious I must remind the reader that the more abiataxic and apolemitic is a society the less does it reflect the features of its military organization.

Tallenic military organization is characterized by high M.P.R., low cohesion and low subordination. We should expect to find it, therefore, in egalitarian societies with rather inarticulate, amorphous political organization. High M.P.R. means that there is no warrior stratum, having the monopoly of arms, and thus able to acquire privileges. Nor are there, in view of the low degree of subordination, any rights to command which would enable their holders to elevate themselves above the common crowd. Indeed, any cohesion that there is in such societies stems from either far-reaching kinship links (e.g. the Tallensi) or from the consciousness of cultural sameness (e.g. North American frontiersmen or the Trekboers of South Africa); or from both, plus some other factors. This type of military organization can exist only where there is no real government, because the first task of any effective government is to organize defence. It is evident, therefore, that it can exist in an adulterated form only in small societies, because an integration of large numbers requires an organized government. Nor can it survive in conditions of intensive and frequent warfare because under such circumstances leadership and discipline are indispensable. The case of the American frontiersmen seems to contradict this statement, but not on closer inspection. True, they waged war on the Indians almost constantly; but because of their very substantial technical and numerical superiority they were not compelled to brace themselves into a tightly knit body—as were the Cossacks, their equivalents on the Eurasian steppe. With very few exceptions, which can be accounted for by equally peculiar circumstances, this type of military organization is found only in small, isolated tribes, usually situated in inaccessible places, whether mountains or deserts (e.g. the Tallensi, the Eskimos, the Australian aborigines). The lack of stratification and the political amorphism, which are the usual features of such societies, also characterized the trans-Appalachian states of America in the early nineteenth century. There, everybody had his gun, and the regular troops were very weak; this was one of the most important conditions moulding the Frontier society, and gave a singular twist to the whole American civilization; its repercussions reverberate to this day. The almost complete fluidity of social structure, the extraordinary impotence of the central government, the rugged individualism, the extreme egalitarianism, the sheriff's rough justice and lynchings could not have survived under a government possessing a large disciplined army and a well-organized administrative machine; both of which would have been indispensable had the United States been exposed continuously to a serious external threat.

The masaic type of military organization differs from the tallenic only in its greater cohesion; M.P.R. remaining high and the degree of subordination low. It is usually found in societies which are small and unstratified, but possess definite political organization, even though this organization is extremely democratic. Its cohesion is mainly the result of military cohesion necessitated by constant warfare. Clearly, this type cannot exist in large societies because the strong cohesion of large armed forces can only be achieved through subordination. Geographically, societies whose military organization is of this type are situated in easily accessible places, such as open steppes or the coasts (e.g. the Masai, the Cossacks before they were incorporated into the Russian army, the Cretan pirates in antiquity). The enumeration of all such societies would be very tedious, because practically all small and warlike societies, be they Eurasian, African or Amerindian, belong to this category. They can have leaders, who are, however, as a rule of charismatic type, and whose authority is very circumscribed: because of the small size of the population no authority able to keep down the majority can establish itself. If there is such an authority, it is invariably that of a priest, and therefore unconnected with military organization. Naturally, there are many societies which approximate imperfectly to this type, or where the coalescence of the functions of the priest and the war leader makes it difficult to classify them (e.g. a number of Bantu tribes of southern Africa).

The mortazic type of military organization is characterized by low M.P.R., high cohesion and high subordination. It is not surprising, therefore, that it is found in societies which are steeply stratified and monocratically ruled. The cohesion of these societies may, but need not, be deep; it may embrace the ruling stratum only, while the rest of the population may be living in well-nigh self-sufficient and isolated village-communities; that is to say, the organization of these societies may be segmentary (e.g. the Kitara Kingdom in Central Africa). This type of military organization can exist in very large societies, like the Roman Empire or Manchu China, but also in very small ones, like Milan under the Sforzas and other cities of Renaissance Italy ruled by the condottieri; but generally it is found in societies larger than those whose military organization is of tallenic or masaic type; it is not, therefore, surprising that the former exhibit greater variety than the latter.

Societies whose military organization is of the mortazic type can be divided, in the first place, into two main groups: (1) those whose cohesion has also other than military roots, such as economic interdependence or national consciousness (e.g. France under Louis XIV or Japan under Hydeoshi); and (2) those whose cohesion resides almost solely, or at least primarily, in the army (e.g. the Abbaside Caliphate which owed its existence to its mercenaries—mortazeh, as they were called).

Secondly, these societies could be classified into those where the armed forces and the ruling stratum coincide (e.g. the Ottoman Empire), and those where they do not. Among the latter there are two varieties: (1) all warriors may belong to the privileged strata, which, however, contain other kinds of people as well, such as the priests or the literati (e.g. Ramseside Egypt); (2) only officers belong to the privileged strata, while the rank and file are drawn from unprivileged strata (e.g. European monarchies of the 'Enlightenment'). As has been already indicated, the position of the ruling class is less secure in the second situation, which, moreover, can exist only in imperfectly biataxic societies, that is to say, societies where the distribution of ophelimities is not determined by naked force alone. Interstratic mobility tends to be high in societies whose military organization is of mortazic type; in some it was indeed extreme—in the Ottoman Empire, for instance, or the Mughul Empire or the Abbaside Caliphate. Some further illustrations of this point can be found in my article in *Social Forces* (October 1950).

Other things being equal, the extent of governmental regulation tends, as we saw earlier, to co-vary with M.P.R. It is understandable, therefore, that in societies possessing mortazic type of military organization, its range is rather restricted. This is not, however, an absolute rule because the extent of governmental regulation is influenced by non-military factors too. The political structure tends to be that of an autocracy, but of a superficial, non-penetrating kind; let us call it crust-autocracy. The Islamic states come under that category almost without exception.

The homoic type of military organization is characterized by low M.P.R., high cohesion and low subordination. Typically, it is found in nobiliary republics marked by steep stratification, high cohesion and egalitarianism within the ruling stratum. The first is connected with low M.P.R., while the other two are reflections of the corresponding features of military organization. This type of military organization cannot be found in anything approaching its pure form in large societies, because high cohesion and lack of hierarchization can be reconciled only if the numbers involved are small. In large societies, particularly when the means of communication are undeveloped, any flattening of the pyramid of subordination must mean at least partial political disintegration. Understandably therefore, this type of military organization has thrived only in a type of society which may be called the nobiliary polis, where privileged warriors lived in propinquity and dominated their serfs collectively, and of which Sparta and the Dorian state in Crete are most representative.

The ritterian type of military organization is marked by a low M.P.R., low cohesion and low subordination. The political form which usually accompanies it is that of a nobiliary republic, even if it is a monarchy nominally. But, in contrast to the nobiliary polis, here the nobles are strewn over a wide area, and dominate their subjects individually, not collectively. Let us call such a system a seignorial republic. The society as a whole is characterized by steep stratification, connected with low M.P.R., low cohesion and egalitarianism within the ruling stratum. The purest examples of such a society are: Poland and Hungary at the end of the Middle Ages, Parts of Germany dominated by the 'free knights' during the same era, the Radjput states in India around the thirteenth century, and the Parthian Kingdom; above all, Poland. It would be inappropriate to describe these societies as feudal, because in its strict sense the term feudal pertains only to the juridical relations arising from the bond of vassalage. True, it is also a matter of discipline, which within a feudal hierarchy is as a rule much laxer than within a bureaucracy; and if it is strengthened beyond a certain point, the feudal state becomes an autocratic bureaucracy, and its military organization mortazic (or neferic if M.P.R. is raised at the same time); if the discipline is relaxed so much that the hierarchy practically dissolves itself, the state becomes a seignorial republic, and its military organization ritterian, the ultimate logical end being complete pulverization. Feudalism, therefore, is a transition zone between the extreme poles of political organization—centralized bureaucracy and the multitude of semi-independent domains; and in its military aspects, it stands between the mortazic and the ritterian types of military organization.

The relationships between the warrior nobility and their subjects, and their juridical claims to the soil, may assume diverse forms. The land may be the allodial property of the nobles (e.g. Poland and Hungary), or it may be held as fiefs (e.g. western Europe), or as mere beneficia (e.g. Muscovy and Mughul India). The peasants may live in self-governing villages and be only required to deliver dues, though these may be very high (e.g. India); or may be incorporated into manors and obliged to till the lords' demesnes besides their own fields (e.g. the medieval Occident); or they may be mere labourers, or even slaves (e.g. colonial Brazil). The extent of the lords' jurisdiction varies greatly too. As far as military organization is concerned, these variations are relevant only in so far as they affect the inner structure of the warrior stratum; and here a rough inverse correlation can be discerned between the extent of the nobles' jurisdiction and the security of their tenure of land, on the one side, and their subordination to the ruler, on the other. Flattening of the feudal pyramid usually means the transformation of fiefs into allodial property.

The term feudal is sometimes twisted to designate simply the exploitation of the masses. I shall not waste time in discussing the merits, or rather demerits, of such an abuse of words. But it even seems inappropriate to apply this term to all societies where a warrior stratum, possessing the monopoly of arms, rules unarmed masses. The habit of twisting the meaning of old words in order to express new concepts is very pernicious because it creates confusion; it is much better to invent new words. And as the type of society where a stratum of warriors dominates unarmed masses is certainly common enough to deserve a name, I propose to call it bookayan, and the dominating warrior stratum a bookay. These terms are derived from buke—the Japanese word for the military nobility. A bookayan society may be a nobiliary polis, a seignorial republic or an absolute monarchy, and its military organization may be mortazic, homoic or ritterian.

The ritterian type of military organization is usually found in larger societies than the homoic type, because when large numbers are involved, low subordination produces low cohesion.

The neferic type of military organization is marked by high M.P.R., high cohesion and high subordination. It is generally found in societies possessing the following features: (1) a tendency towards the levelling of social inequalities, connected with high M.P.R. and high subordination, counterbalanced by (2) stratificatory tendencies produced by high subordination; (3) a tendency towards totalitarianism connected with high M.P.R.; (4) monocratic government; (5) high cohesion which, however, may be merely a reflection of the cohesion of the politico-military machine (e.g. the Inca Empire), or may also be rooted in ethnic consciousness and/or economic interdependence (e.g. Nazi Germany); (6) high interstratic mobility, a correlative of monocracy. The purest example of such a society is the Soviet Empire; and undoubtedly, totalitarian, bureaucratic despotism is most congenial to this type of military organization. All societies of the past which possessed it fall into this category: the Ch'in kingdom, China under the Han and T'ang dynasties, the Inca Empire, the Old Kingdom of Ancient Egypt, Japan in the Taikwa period, Siam in the Ayudhya period. And of these the two whose military organization was most purely neferic—the Ch'in and the Inca states— were also the most despotic and totalitarian.

It may be objected that in modern times societies whose political constitutions were democratic and liberal possessed this type of military organization. Before I answer this objection, I must remind the reader how recent, imperfect and fragile is this institutional arrangement described as liberal democracy. The adjective 'liberal', I must add, designates the reluctance on the part of the state to infringe upon certain spheres in the life of its citizens, considered as their liberties. It must be remembered that, although in political polemics democracy and liberalism are confounded, they are by no means identical. A majority rule can be very tyrannical and show complete disregard for the freedoms of the subject. Such were the democracies of ancient Greece, to which the notion of inalienable freedoms was unknown. The extension of military service may have in some ways fostered democracy; it certainly produced some levelling of political rights; but there is not the slightest reason to believe that it was favourable to liberalism, the growth of which was due to other causes. But to revert to democracy as such, the astonishing thing is the novelty of this regime as far as large states are concerned. Among these the United States, in spite of their youth, have possessed a democratic constitution for the longest time. The British constitution did not become democratic even in principle until the eighties of the last century. France became a democracy only with the establishment of the Third Republic; the previous attempts were abortive. The Italian state was parliamentary from its inception because it owed its birth to popular revolts, which paved the road for the house of Savoy; but it was soaked in patronalism and graft, and succumbed to Mussolini after fifty years of existence. The Weimar constitution was notoriously short-lived. In Poland and some other eastern European countries parliamentary, democratic constitutions operated for only a few years: in Russia, never. Czechoslovakia was exceptional. There, the lack of a native nobility (which had been exterminated three centuries earlier by the Germans) made the Czech mentality peculiarly immune from the worship of the sword and the traditions of lordly haughtiness; and it attenuated class animosities within the Czech people as the landlord or the capitalist was mostly a national enemy too. Also, in contrast to other countries of eastern Europe, Czechoslovakia possessed an indigenous middle class, usually the main pillar of parliamentarianism. Moreover, being exceptionally prosperous it was not, like its neighbours, afflicted with hosts of desperados. This case shows that military factors do not determine everything. Nevertheless, it is significant that Czechoslovakia was very unmilitaristic, and that other examples of working democracies, and the best working at that, are provided by countries which were demilitarized and neutral: Switzerland, Holland and the Scandinavian states.

The democratic constitution of the United States was shaped, as I pointed out before, when their military organization was essentially tallenic. Britain was practically demilitarized, and it was precisely the unmilitary character of the ruling class which compelled it to compromise with the proletariat. Neferic military organization was adopted by these countries only half-heartedly during World War I, and it is only since World War II that it may have affected their political life. The French republic is the only state where neferic military organization coexisted with parliamentary democracy for a fairly long time. But even this case does not prove their congeniality. In spite of general prosperity, complete lack of population pressure, and ethnic homogeneity, all providing a propitious setting for an abiataxic political system, the gravest menace to the republic did not, until the rise of the Communist Party, stem from proletarian movements but from the army; the Bonapartist ideal of an egalitarian dictatorship exercised a great deal of attraction; and it seems that only lack of nerve prevented General Boulanger from establishing himself as a dictator. Even in this case, therefore, the neferic military organization tended to introject its features into the social structure, in spite of the fact that this was the most peaceful period in the history of Europe, and that demographic factors made the French nation unaggressive.

The most important proof, however, that neferic military organization tends to cast a society into a totalitarian and autocratic mould, is the fact that all states which participated in the two World Wars moved in this direction irrespectively of their traditional ideals; and after the emergencies remained nearer to this type than they were before. Moreover, and this is the most startling proof of all, the Soviet Empire was the only state whose structure remained completely unaffected by the last war and the 'cold war' which followed. This shows beyond any doubt that its structure must fit perfectly the needs of modern warfare.

DETERMINANTS OF TRANSFORMATIONS

All human arrangements undergo constant change and, therefore, having propounded a classification of the forms of military organization, we must deal with the problem of transitions from one type to another. But before we examine various modes of transition, we must attempt to recapitulate the question of the factors which produce these transitions. Although all the preceding parts consisted of the discussion of these factors, we must try now to connect it with our general classification; and as this is based on three variables we must establish the factors which determine their values.

M.P.R. is determined by the following factors:

(1) The optimum M.P.R., which, it will be recalled, depends on various circumstances affecting the technique of war, the most important being the relation between the cost of the most efficient armament and productive capacity.

(2) The tendency towards the full utilization of manpower, whose strength depends on the exigency of military effort, which in turn depends on the seriousness of the threat from outside and on the strength of internally generated bellicosity.

(3) The tendency on the part of the ruling stratum to monopolize arms in order to fortify its position. If the optimum M.P.R. is high, this tendency stands in opposition to the preceding. Their interplay is connected with intra-hierarchic oscillations, the ruler favouring the extension of military service as a means of strengthening the army and of curtailing the privileges of the warrior nobility at the same time. Where there is no stratification, the tendency towards internal conquest operates.

(4) The tendency to conquer, which, if successful, must lower M.P.R., because it means subjugation of hostile peoples; there are exceptions, however, where the conquered are incorporated and used militarily. But success in a career of conquest may require an approximation to the optimum M.P.R., which may be high, in the original political unit.

The degree of subordination is determined by the following factors:

(1) The intensity of warfare, which fosters it.

(2) The size of the armed forces: the bigger they are the more indispensable is a high degree of subordination.

(3) The availability of the means of control, such as the police and administrative technique, etc.

(4) The tightness of the ruler's control over equipment and remuneration.

(5) The facility of suppression, which in its territorial aspect depends on the attack-versus-defence relation.

(6) The constellation of power centres, which may have a positive or negative influence on the degree of subordination. We shall say that the value of its influence increases when it passes from the negative to the positive.

The degree of cohesion is determined mainly by the following factors:

(1) The pressure from outside, which tends to heighten it (comp. Seeley's 'law').

(2) The imperativeness of co-operation, which depends on technico-tactical circumstances, which may make it a more or less absolute condition of victory.

(3) The pheric size of the unit; other things being equal, the smaller it is the more cohesive it tends to be.

(4) The degree of subordination, which implies cohesion (but not vice versa). This cannot be considered as an independent variable because it is determined by factors already indicated. It follows that all factors influencing subordination affect cohesion too.

(5) The economic interdependence, cultural homogeneity and other non-military binding circumstances of the society.

(6) The pressure from below on the bookay (the ruling warrior stratum), which may assure its inner solidarity. This applies, of course, only to bookayan societies.

TYPES OF TRANSITION

As there are six types of military organization, there must be thirty types of transition from one type to another. These, it must not be forgotten, are pure types of transition, in the same way as the static types are pure: none of them occurs in reality without an admixture of traits belonging to another type. It follows that the examples given present only the nearest approximations to the pure types which I could find. The vast majority of changes in military organization which have occurred, or are occurring, cannot be put into any of these categories because they contain traits of several of them at the same time. Nevertheless, this classification is useful as an analytic tool, because it enables us to order the bewildering diversity of the reality, and to describe any actual case in terms of the degrees of admixture of a limited number of elements.

The following are the pure types of transition:

I. Tallenic (Msc) to Ritterian (msc)

I could find no example of this type, and there are reasons why it is unlikely that it ever materialized. The lowering of M.P.R. can occur (1) as a result of a conquest from outside or from inside (e.g. by a party); or (2) in consequence of the lowering of the optimum M.P.R. under circumstances compelling the adoption of the most effective military organization; or (3) because of the voluntary disarmament of the bulk of the population. Now, the possibilities (1) and (2) imply a condition of severe struggle, which would undoubtedly raise the degrees of subordination and cohesion, thus leading to the mortazic type. Possibility (3) could materialize only in a large pacified state, secure from external danger, whose existence is incompatible with either tallenic or ritterian military organization.

II. Tallenic (Msc) to Masaic (MsC)

The variable involved here is the degree of cohesion, which is raised as a rule by the intensification of warfare, or by the advent of circumstances enhancing the imperativeness of co-ordination.

E.g.: the Cossacks of the Dnieper were originally an agglomeration of bands of run-away serfs and criminals, roaming over the wild steppes. As pressure on them from the surrounding states grew, they organized themselves, and came to constitute a cohesive body. The intensification of warfare among the Amerindians of the Northern Plains, consequent upon the introduction of the horse, made their tribal organization tighter.

III. Neferic (MSC) to Mortazic (mSC)

This type of transition may occur in consequence of a conquest, after which the conquered are disarmed and the conquerors reserve for themselves the right and duty of military service; e.g. the Mongol conquest of China. But the lowering of M.P.R. may also be due to a change in the optimum M.P.R., which makes military utilization of the whole manhood unprofitable; this is what happened in Egypt and Israel after war-chariots came into use.

IV. Masaic (MsC) to Homoic (msC)

I could find no example of this type. The reasons why it is unlikely that it has occurred often are similar to those indicated under I. The resistance of the existing social structure to the lowering of M.P.R. can be overcome generally only under circumstances requiring a strenuous military effort, which would favour subordination; whilst a conquest, whether from outside or inside, would promote subordination too.

V. Masaic (MsC) to Neferic (MSC)

The increase in the degree of subordination can be due to many factors, but in a primitive society already engaged in intensive warfare it is, as a rule, the result of numerical growth. This growth could be a simple proliferation, but if all the land is occupied, that can seldom go very far. Conquest, therefore, is the usual way of growing. But conquest can lead to the establishment of neferic military organization only if the conquered are incorporated into the body politic of the conquerors. If they are turned into serfs or slaves they must be disarmed. The erection of the Zulu kingdom and of several kingdoms of Eurasian nomads, as well as that of the Almohad state in Morocco, represent this type of transformation.

VI. Masaic (MsC) to Tallenic (Msc)

The loss of cohesion of the military organization, which is the essence of this type of transformation, and which tends to produce political disintegration as well, is usually due to the abatement of warfare; e.g. the Berber republics in North Africa after the pacification of the country by the French, various Bantu tribes after their subjugation by the colonizing powers, the Chuckchees after their migration from the steppe into the tundra.

VII. Mortazic (mSC) to Neferic (MSC)

Here, the extension of military service may be due to the rise of the optimum M.P.R., combined with the pressure of warfare; e.g. European states between 1750 and 1950. This type of transition tends to promote the change over from a crust-autocracy to totalitarian autocracy, because the rise in M.P.R. tends to produce the extension of the sphere of governmental regulation.

VIII. Mortazic (mSC) to Homoic (msC)

This type of transition may be the consequence of changes in the economy, or in the constellation of power centres, or of the abatement of warfare, affecting adversely the strength of the ruler. It can take place fully only in small states, because only where small numbers are involved is low subordination compatible with high cohesion; e.g. the development of the Spartan *de facto* republic. In large states this type of transition can proceed only half way—to pretorianism (e.g. the development of the Mamluk state in Egypt); because any further lowering of subordination would produce the loss of cohesion and the resultant type would be ritterian.

IX. Homoic (msC) to Masaic (MsC)

Such transition may take place as a result of the rise in the optimum M.P.R.; e.g. the transformation of the Athenian military organization during the Persian Wars—the creation of a large navy in which even the poorest could serve—which prompted the establishment of the egalitarian sailors' republic. This example is, of course, imperfect because no account is taken of the slaves and the metics.

X. Tallenic (Msc) to Neferic (MSC)

The only cases of direct transition of this type, which I could find, are those where some tribes were incorporated into a state possessing neferic military organization; e.g. the Inca and Zulu conquests.

XI. Tallenic (Msc) to Homoic (msC)

An example of the transition of this type is provided by the settlement of the Central-Asiatic tribes (ancestors of the Radjputs) in India. Their tightly organized clans established themselves in military colonies amidst loosely organized peasant populations, which they subjugated; the resultant societies strongly resembled Dorian cities. In this case the drop in M.P.R. was due to conquest, whilst the struggle and the necessary preparedness for combat assured cohesion.

XII. Masaic (MsC) to Mortazic (mSC)
XIII. Tallenic (Msc) to Mortazic (mSC)

These very common types of transition are the correlatives of the process of state-building through conquest, which naturally results in the lowering of M.P.R., while the rise in subordination is due both to the warfare preceding the conquest and to its aftermath, viz. the enlargement of the political unit and the need for readiness to suppress revolts; e.g. the establishment of various kingdoms in the Sudan and East Africa, and of Berber principalities in southern Morocco. These are the most recent among a multitude of similar cases.

XIV. Mortazic (mSC) to Masaic (MsC)
XV. Mortazic (mSC) to Tallenic (Msc)

These types are correlatives of the process of disintegration of primary conquest kingdoms, which resolve themselves back into tribal units. This is what happened to a number of Sudanese kingdoms. Various peoples surrounding the nucleus of the Bornu sultanate in Sudan underwent such transformations, and the reverse, several times within the recorded history. The most grandiose examples are the downfalls of the Khmer Empire in what is now Indochina, and of the Almohad state in North Africa.

XVI. *Neferic (MSC) to Ritterian (msc)*

The purest example is the development of feudalism in Japan. There, under the circumstances of absence of external danger, cohesion and subordination declined. A stratum of warriors differentiated itself and subjugated the rest of the population; thus the M.P.R. was lowered.

XVII. *Ritterian (msc) to Neferic (MSC)*

This type is a correlative of the process of replacement of 'feudalism' by a centralized state based on universal conscription. This process may be a result of changes strengthening the position of the ruler and of the intensification of warfare, under the condition of high optimum M.P.R.; e.g. the reforms of Shang Yang in the Ch'in kingdom in the fourth century B.C., and of Mehmet Ali in Egypt in the early nineteenth century.

XVIII. *Neferic (MSC) to Masaic (MsC)*

This type of transition cannot occur in a large state because masaic military organization is incompatible with largeness. Neferic military organization, on the other hand, is not found in very small societies because such societies are never durably monocratic. This type, therefore, is the correlative of political disintegration; e.g. the break up of various nomad kingdoms in central Asia.

XIX. *Neferic (MSC) to Tallenic (Msc)*

What has been said about the preceding type applies to this one too, except that here, disintegration needs to proceed even further, as it did among those of the above-mentioned peoples which were driven from the steppe into the forest or the tundra, and abandoned the warrior-pastoralist's way of life.

XX. Homoic (msC) to Tallenic (Msc)

This type is very rare because factors which might raise the M.P.R. (intensive warfare and monocracy) would not permit the drop in cohesion. Moreover, a polity possessing homoic military organization, being necessarily small, is not likely to dissolve itself. Only the destruction of a polity, an invasion or a revolt of the serfs, can bring about this change, which means a reversion to primitive tribalism; I could find no well-authenticated example.

XXI. Ritterian (msc) to Tallenic (Msc)

Ritterian military organization is found only in fairly large states, while tallenic only in small ones; such transition, therefore, could take place only as an accompaniment of political disintegration. But, as such a process is usually coupled with severe struggles, it could hardly produce a form of military organization marked by low cohesion; a more probable outcome would be the establishment of the masaic type. This would give us the following type of transition.

XXII. Ritterian (msc) to Masaic (MsC)

Even this type is very rare because the usual end of large states is to be conquered by other large states: spontaneous disintegration is relatively infrequent, and as a rule limited to primary conquest states. Ritterian military organization, moreover, is rare in a form approaching purity because it thrives only under circumstances of unintensive warfare, and only seldom can a large state find itself in such a position.

XXIII. Ritterian (msc) to Mortazic (mSC)

This type occurs when obedient professional troops replace a feudal levy or a nobles' militia; it is promoted by all factors which foster centralization; e.g. the displacement of feudalism by professional armies and absolute monarchies in western Europe.

XXIV. Mortazic (mSC) to Ritterian (msc)

This type is a correlative of the process of 'feudalization' of conquest states. Once the warriors are dispersed over a wide area, and particularly after ethnic amalgamation with the conquered, they lose their *esprit de corps* and, having gained the allegiance of their subjects, tend to emancipate themselves from the ruler's control. A classic example of such a process is the feudalization of the Carolingian successor states.

XXV. Masaic (MsC) to Ritterian (msc)

Such a transition is unlikely to occur for the reasons indicated under I. No case observed. The usual intermediary stage is the mortazic type. The nearest approach was the feudalization of northern Germany by Charlemagne.

XXVI. Homoic (msC) to Mortazic (mSC)

This type of transition is a correlative of a transformation of an oligarchy into a monocracy, which may even be due to non-military factors: e.g. the rise of tyrants in various Greek cities, or the transformation of the nobiliary republics which existed in northern India into kingdoms during the Mauryan era.

XXVII. Ritterian (msc) to Homoic (msC)

No case observed; if there were any, they were probably accompanied by political disintegration, as no increase in cohesion of a large body is likely to take place without a corresponding rise in subordination.

XXVIII. Homoic (msC) to Neferic (MSC)

This may be a consequence of the tendency, already discussed, towards an alliance between an autocrat, or an aspiring autocrat, and the populace, whom he helps against the oligarchs but wants to use militarily; e.g. the reforms of Cleomenes in Sparta in the third century B.C.

XXIX. Neferic (MSC) to Homoic (msC)

This transition, the reverse of the preceding, may result from a successful curbing of the authority of the ruler by the oligarchs, who monopolize arms; e.g. the events in Sparta after the downfall of Cleomenes.

XXX. Homoic (msC) to Ritterian (msc)

The purest example I could find is the decentralization of the Mamluk state.

Not all of the types of transition discussed above are of equal importance and frequency. They are not, moreover, equally reversible. As noted previously, the tallenic, masaic and homoic types of military organization are compatible only with small political units, except under quite extraordinary circumstances. In view of the general trend in the evolution of societies towards largeness and complexity, it is natural that the cases of transition from any of the types just mentioned to the other three, which are compatible with largeness, are more frequent than the reverse processes. In order to simplify the following sentences, let us call the types of military organization which are compatible with largeness—the types of higher order, and those which are not—the types of lower order. Now, states possessing military organization of a higher order type are just as liable to be destroyed as those whose military organization is of a lower order type; but they are usually swallowed by their peers. A wholesale disintegration and reversion to simpler forms is very rare and, almost without exception, only temporary.

The following types of transition materialize most frequently: tallenic to mortazic, and masaic to mortazic, which are correlatives of the process of primary state building through conquest; the reverse process of disintegration of such states is fairly, but not equally, frequent. Transitions from mortazic to neferic types, and vice versa, are also very frequent, and mostly due to technico-tactical factors; they may well occur in the future. The movements between mortazic, or mortazico-neferic, and ritterio-homoic forms are typical correlatives of oscillations between monocratic centralization and 'feudalism', which are a perennial feature in the life of all large agrarian states.

XI

Revolutions

THE present chapter is not intended to be a comprehensive analysis of the problem of revolutions. Such an analysis will be given in a work on Revolutions, their Types, Causes and Effects, which is in preparation. Here I shall limit myself to a few preliminary remarks.

Before enquiring into the connections between forms of military organizations and revolutions, it is necessary to place the problem of revolutions in its proper setting. As usual, we must begin with a definition. By 'revolution' I mean a violent and illegitimate overthrow of authority. By using the word 'illegitimate' I do not wish to convey any deprecation of such a process on my part, but I merely wish to distinguish it from a violent overthrow of an existing authority which is sanctioned by law or custom. In olden times when the king of Kitara fell ill, and did not commit suicide, it was incumbent upon his sons to kill him, lest a calamity befalls the kingdom. Such an overthrow of authority would not be covered by the term as defined above. Naturally, the limits of the applicability of this, as of any other sociological term, cannot be demarcated clearly. At which point does an authority become established? Where should we draw a line between war and revolution? In Tonga, for instance, a tribe which had beaten all other tribes used to attain a customarily entrenched supremacy—not, let it be noted, a mere de facto domination—which it maintained until the verdict by battle went against it. Should such a supremacy be called a hegemony or an authority; and were the fights which brought about shifts of supremacy, and which were declared and fought according to definite rules, wars or revolutions? Similar questions could be asked in connection with events such as the overthrow of the Assyrian Empire by its vassals, or many episodes in the struggles between Guelfs and Ghibellines. So, here again, we must direct our attention to the clear-cut cases in the first place.

The question why a revolution took place must be split into two: first, why did the rebellion occur, and secondly, why did it succeed? It must not be assumed that there is a simple relation between the incidence of rebellions and their success. There have been several states, such as Togukawa Japan or the Liao Empire in Northern China, where rebellions were almost seasonal events, but all were suppressed. On the other hand, in France, three out of four serious rebellions which broke out during the hundred and fifty years preceding the first World War turned into revolutions.

The incidence of rebellions depends to a large extent on biataxy, but it is also influenced by expectations, which may stand in no relation to objective chances of success. For this reason ideologies instilling the hope of a victorious rebellion or the opposite, are very important. The ideological factors largely determine whether a revolution is followed by a transformation of social structure, or whether its only result is a reshuffle of the personnel. At this point I must remark that it is woefully confusing to use the word 'revolution' to designate any radical transformation.

The incidence of rebellions largely depends on biataxy, but their success in a biataxic society depends mainly on the balance of strength between the supporters and the enemies of the existing order. In order to single out factors which determine this balance we should find answers to a number of questions, among which are the following: On what does the cohesion and vigour of the ruling group depend? On what does the existence of concord depend? What are the circumstances under which a subversive movement can be organized? These and other equally interesting questions cannot be answered in this work, with the result that the exposition which follows is rather unsatisfactory.

The form of military organization determines to a very large extent the distribution of power, which in turn circumscribes the ability to revolt successfully. Now, as a revolution can only take place in a stratified society, we can leave out of consideration societies whose military organization is tallenic or masaic, because these types of military organization are found only where stratification is rudimentary. Speaking approximately, without the pretence of exact quantification, we can say that mass rebellions— that is to say, rebellions carried out by considerable portions of unprivileged strata—are as frequent in societies where M.P.R. is low as where it is high; they appear to have been as frequent in medieval Europe or Japan as in the Chinese Empire. Their outcome, however, is influenced by M.P.R. As I pointed out in an earlier chapter, several peasant rebellions succeeded in China whereas they all failed in medieval Europe and Japan. I have also shown that the revolutions in modern Russia, Germany and Austria took place largely in consequence of the introduction of universal conscription; otherwise as in the past, a defeat in war would not have led to the collapse of the established stratification.

In conditions of low M.P.R. the outcome of an uprising of the lower strata depends on whether the armed forces and the ruling stratum coincide or not. If they do not, then, such an uprising can succeed because, as can be seen from the example of the French Revolution, the troops drawn from the lower strata are likely to join the rebels. Where the armed forces constitute the ruling stratum—that is to say, in a bookayan society—a revolt of the lower strata cannot succeed unless the bookay is either very weak numerically or torn by dissension or quite ineffectual. When the Mongols were expelled from China by a mass revolt they were characterized by all of these attributes.

The provenance of the rebel leaders who become a new ruling group is a very important feature, distinguishing various kinds of revolutions. A revolt within the upper stratum through which one clique supplants another is quite a different thing from a revolt which leads to the replacement of a ruling stratum by another, drawn from among the lower strata. The latter is necessarily a much more profound upheaval than the former. It must not be imagined, however, that a supplanting of a ruling stratum must inevitably produce a thorough transformation of social structure— several Chinese revolutions are proof that it need not be so. Now, the chances of the success of a rebellion depend on whether the rebels possess arms and the knowledge of how to use them. It follows that the higher the M.P.R. the larger is the circle from which the rebels are likely to come. This is why revolutions occurring in bookayan societies are as a rule of the nature of palace revolutions, through which one section of the bookay supplants another at the top of the social pyramid. Japan was the land par excellence of revolutions of this type. On the other hand, China, whose military organization approached most of the time the neferic type, was the state which experienced the greatest number of revolutionary supplantations of whole ruling strata by others drawn from among the lowly.

It may be objected that there are many examples of revolutions which occurred in states possessing mortazic military organization, through which men of very lowly origins attained supreme dignities. A large number of such examples can be found in the history of the Late Roman Empire, the Byzantine Empire and the Islamic Orient. The Delhi sultanate even had a Slave Dynasty. Several Mamluk sultans started their careers as slaves. All this is undoubtedly true but it must be remembered that these people did not revolt when they still belonged to the unprivileged strata; nor was their following drawn therefrom. They were drafted into the armed forces, and raised to positions of authority by the favour of the ruler, and only then did they rebel with the aid of their professional soldiers. Pretorianist revolution is the type which as a rule occurs under mortazic military organization.

Rebellions which occur in states whose military organization is ritterian are mostly such that it is difficult to decide whether they should be called revolutions or not. In feudal and seignorial, semi-feudal societies, which are the ones that possess this type of military organization; most rebellions were attempts to gain territorial independence—not to usurp the supreme authority—owing principally to the low cohesion of these societies and their military organization. For such uprisings we might reserve the term 'insurrection'. The dividing line between an insurrection and a revolution is, of course, blurred. Furthermore, certain kinds of rebellion may become legalized, and, therefore, cease to be rebellions. Thus, for instance, in the Polish kingdom it became a custom towards the end of the Jagellon dynasty that nobles, before embarking upon a campaign, would bargain with the king and try to extort various privileges.

Generally speaking, a fairly biataxic society is stable only if there is a congruence of the distribution of real military power with the distribution of wealth and political rights. It follows that any change in the distribution of military power is likely to lead to a revolution provided that:

(1) A society is biataxic.

(2) The distribution of political rights and wealth has a certain inertia of its own, and does not adjust itself automatically.

XII

Concluding Remarks

WE are approaching the end of this investigation. This does not mean that the questions are exhausted. Scientific investigation is like a pursuit of the horizon: every step opens new perspectives, a solution of any problem always raises a host of new questions. I have traced here relations between groups of circumstances, construed into variables. The selection of them has been guided solely by the aims of the present investigation, and is by no means intended to impart to these variables any sort of universal preponderance. Such monism would be as nonsensical as economic or any other monism. We must never forget that all our factors are mental constructs, and in particular, that terms like economic, political, religious, technical, military, etc., are only labels. In view of the uncertainty of boundaries, and of overlappings, between their fields of applicability, it is altogether absurd to make any of them into a prime mover of all social change.

The relation between the cost of armament and productive capacity, to take one instance, is one of the determinants of military organization. In turn it is determined by other factors, such as, for instance, technical progress. And naturally one is led to ask: on what does that depend, and how is it connected with military circumstances? I was particularly tempted to give some answers to this point, because the present book grew from a chapter of a work on the social conditions of technical progress which I have been preparing for a number of years; but in view of the complexity of the problem I must confine myself to the following brief remarks.

The elimination of technically backward societies was one of the principal mechanisms through which technical progress was accomplished. Also, the growth in size which was essential to this progress was accomplished through war and conquest. These, furthermore, were instrumental in bringing into existence the leisure class, without which science could not have arisen. Political rivalry and warfare, moreover, are the circumstances which most often prompt rulers to promote technical progress. But on the other hand, though success in war undoubtedly depends on technical skill, in the past this skill was mostly traditional, often enveloped in magic, and acquired without assimilating the mentality which produced it, whilst discipline, *esprit de corps* and savage bravery were more immediately effective in assuring military ascendancy. And the habit of unquestioning obedience is in many ways incompatible with the critical, reasoning attitude, essential to inventiveness. During the era of the ascendancy of the nomads, which began with the invention of the stirrup and ended with the invention of fire-arms, the military advantage lay with a simpler type of society, which entered a technical blind alley. Furthermore, subjugation produces 'parastic appropriation of surplus' (for the definition see the glossary), which is one of the most powerful among factors inhibiting technical progress.

These remarks are just a sample of the ramifications of the present enquiry which could be tracked. Another task would be the retesting of the generalizations propounded, and, above all, the quantification of variables. In principle such quantification is possible because nearly all concepts describing structural varieties are polar; so that if some method of measuring these variables were devised, the whole theory would be amenable to mathematical treatment. But even without that, further research will lead to its elaboration and modification; particularly as the data concerning non-western societies are not highly reliable, because of the backwardness of historiography. And I hope that some of the theories expounded will be subsumed under wider generalizations yet to be discovered.

The present work resembles a report of a reconnaissance flight which has only succeeded in locating the most important points: the work of topographers is still to be done. Such reconnaissance is absolutely indispensable because in the universe of social phenomena, so fluid and full of diversity, and influenced by so many factors, no investigation of relations between variables, however careful and exact, can be fruitful, unless all the most important factors are taken into account. The statements contained in the preceding pages are derived from a careful study of data. Many of my initial hypotheses have been abandoned or modified, when contradicted by 'brute facts'. Still, many of the generalizations expounded must be described as bold. But it would be a mistake to be too diffident in presenting them, because there can never be too many hypotheses. And if sociology has suffered from excessive theorizing, it was because this theorizing was vague and normative, advancing unverifiable propositions. Every science progressed by careful research guided by bold hypotheses: never by a thoughtless, even though painstaking, heaping up of disjointed bits of information.

DIAGRAMMATIC SUMMARY

Most books very wisely end with a conclusion or a summary; it is certainly advisable to give the reader finally an overall picture. But the present book is so compressed that summary of conclusions would be disproportionately long. Moreover, to sum up the manifold interrelations of variables, one would have to approach the same variable many times from different angles. I decided, therefore, to present a summary in the form of a diagram, which is the only possible way of describing briefly manifold interrelations. The additional advantage is that only a diagram enables one to see simultaneously all variables as parts of one complex constellation. Naturally, such presentation must be even more schematic than the main argument, because many qualifications and elaborations are necessarily omitted.

The meaning of the signs is as follows:

$A \longrightarrow B$ A fosters B, i.e. variations in A tend to produce variations in B in the same direction (i.e. increase or decrease); the reverse is not implied.

$A \dashrightarrow B$ A inhibits B, i.e. variations in A tend to produce variations in B in the opposite direction; the reverse is not implied.

$A \dashrightarrow B$ A influences B in a complex way; the reverse is not implied.

$B \longrightarrow C$ A strengthens the efficacy of B in affecting C.

$B \dashrightarrow C$,,   ,,   ,,

$A \leftarrow \dashrightarrow B$ A and B are structural correlatives.

The diagram is appended at the end of the book.

XIII

A Guess about the Future

GAZING into the inscrutable future is a fascinating though generally unprofitable occupation. And I cannot resist the temptation to advance a guess about the future in the light of the generalizations expounded above. It would be imprudent, however, to predict categorically what will happen; only possibilities can be discussed.

I must point out in this context that the inability to give definite forecasts on all occasions is not merely the result of the backwardness of the social sciences. Even if we were able to estimate very accurately the strength and direction of social forces, the margin of uncertainty would remain, because in many situations the forces tending to push the social structure in opposite directions are so evenly matched that the outcome may be decided by an insignificant unique event.

As far as the future of humanity is concerned, we can discern several possibilities: (1) A balance of power system may continue to operate, producing periodical wars. (2) Wars may lead to a complete collapse of civilization, perhaps to the extinction of civilization, or even the disintegration of the globe. (3) A hegemony of one state, which has succeeded in knocking out all other contestants, may be established. (4) Some kind of world federation may come into existence, assuring universal peace. Let us deal with them consecutively.

It does not seem that any technical or tactical developments are likely to occur which would loosen the cohesiveness of the armed forces. Nothing is likely to bring back an individual manner of fighting, like that of the Homeric heroes or the medieval knights. On the contrary, technical progress steadily increases the premium on organization. The armies of the future are not likely to become less cohesive bodies than they are today. Nor are they likely to become less disciplined, except perhaps at the very top, where, because of the unmanageable technicality of military direction, a committee may replace an individual leader. Otherwise, all improvements in the techniques of organization, communication, detection, etc., will probably strengthen discipline. Only the future state of M.P.R. is open to doubt. This means, in terms of the nomenclature proposed previously, that the military organization of the warring states of the future could be either mortazic or neferic. It is possible in principle that push-button warfare will make large armies useless. Nevertheless, military organization would probably remain basically neferic, because the requirements of such warfare can be met only by mobilizing the whole labour force and turning the country into one great arsenal. The population would thus be militarized even though the majority would not participate directly in fighting. This would be merely an extension of the present trend: even in contemporary armies only a minority of soldiers are directly engaged in killing. It is almost certain that the whole social fabric of these states would be stamped with the pattern of neferic military organization. This would mean that they would all approximate to the type of totalitarian monocracy, of which Soviet Russia is the most perfect example.

These states would not, of course, be exactly similar. But all of them would undoubtedly be characterized by all-pervading governmental control, the concentration of authority in the hands of a supreme ruler or a supreme committee, considerable, but in comparison with many other types of societies relatively slight, social inequalities, and great interstratic mobility. There would be no room in such societies for parliamentary democracy as it has been known in the Occident; nor for the unassailable spheres of freedom of the individual. Even in countries where liberal traditions are deeply rooted, the pressure of circumstances would no doubt prove overwhelming. The world would be covered by garrison-states, constantly preparing for the next bout of wholesale slaughter.

It is evident that the features which constitute the peculiarity of the Occidental civilization could not survive in such an environment. The most prominent of these features is the stress on the freedom of the individual; in other words, liberalism. Liberalism was at the same time the cause and the effect of the ascendancy of the Occident. The cause: because freedom of thought and of enterprise was the chief root of the unprecedented surge of its science and technology. The effect: because its industrial superiority gave it prosperity and security, thus making the ground propitious for the realization of liberal ideals. The inner struggles did not need to be violent (even the French revolution was a comparatively meek affair), nor did the states need to muster all their resources for a life-and-death struggle. The era during which the Occidental civilization decisively differentiated itself from others, and outstripped them in the field of technology, that is to say, between the Thirty Years' War and World War I, was a period of unusually mild warfare, except for the Napoleonic interlude.

The radiations of liberalism into non-Occidental regions was due principally to the military and technical ascendancy of the Occident. Coveting Occidental technical skill, the rulers of Russia, Japan and many other countries thought that they could foster industrialization only by imitating Occidental socio-political institutions. Liberalism attracted them only as a way to military power. The success of the Bolsheviks in building a great industry, able to equip a powerful army, was a turning-point: it showed that totalitarian planning was a much quicker and surer road to military might than economic liberalism.

Birth control and technical progress might enable the countries of Occidental civilization to maintain their standards of living in spite of the loss of their privileged position. But if they have to live in constant preparedness against assaults by enemies over whom they enjoy no technical superiority, they are bound to become thoroughly militarized.

It has been said that liberal regimes are in a stronger position than their totalitarian rivals because, in virtue of the freedom of thought which their citizens enjoy, they will always have a technical superiority. But this argument is far from convincing. Freedom of thought was a condition of the rise of science, when the latter's utility was not evident. But now, when even the most obscurantist government appreciates the utility of natural sciences, scientific research may be fostered even in the most despotic state. There is no reason why research in physics or chemistry should require the freedom to doubt the basic tenets of social creeds; particularly in view of the ability of the human mind to keep bits of knowledge in watertight compartments. A critical and inventive attitude in one's speciality may be accompanied by unbounded credulity in other fields. It is most unlikely, therefore, that this factor could prevent the 'sovietization' of all states under the circumstances of continuous preparedness for war.

This 'sovietization' is all the more probable because many other non-military factors are working in the same direction: above all, economic factors. The growth of economic units and their bureaucratization, combined with the expansion of governmental control, undoubtedly undermine the ramparts of the freedom of the individual. I have already pointed out previously why this process is to a certain extent unavoidable. Nevertheless, though the days of untrammelled capitalism are certainly gone, some sort of corporative society might arise, in which the power of the central bureaucracy might be counter-balanced by that of the political parties, trade unions, manufacturers' and merchants' associations, churches and other bodies. But under conditions of frequent warfare the necessity of unified direction would undoubtedly tilt the balance in favour of the central bureaucracy. Moreover, the suppression of subversive associations, some measure of control over the dissemination of ideas, and active propaganda, are indispensable to the successful prosecution of war, even of the 'cold war'; and all of them pave the road for 'sovietization'. So, in the conditions of modern warfare, all states engaged in it frequently or living in constant preparedness for it must sooner of later become 'sovietized'; even if they proclaim that they are fighting to preserve the freedom of the individual.

This does not mean that the features of the Soviet state which strike Occidentals as peculiarly barbaric and crude would necessarily be taken over. To some extent these features are the inevitable result of extremely totalitarian, monocratic political structure; but largely they are due to the peculiar circumstances under which the regime came into being. In Russia only the upper classes were educated; the masses lived on the level of pre-literate civilization. And once this veneer of upper-class sophistication was removed, the mentality of a primitive people was bound to assert itself. Moreover, one of the reasons why the Hitlerian and Stalinist regimes appear to us so barbaric is that they were built by men who, before they attained power, were regarded as the scum of society. The fascist regime in Italy, in the running of which the pre-existing ruling classes had a much greater share, appears to us infinitely more urbane. This is not, of course, the whole explanation of the peculiarities of these regimes: some of them were undoubtedly due to the fact that these regimes were set up by sectarians. It seems certain that purely military dictatorships would not be so anxious to enthral every nook and cranny of the human mind.

The situation just outlined could hardly persist indefinitely. Constant improvements in the efficacy of weapons and the continuous shrinking of the world in terms of transport and communication would sooner or later bring about either the destruction of civilization (or even of the globe), or the conquest of the whole world by one state. There is no need to contemplate the first alternative. In its milder form it would mean a return to savagery, and it is impossible to foresee along what lines humanity would evolve if it had to begin from scratch once again. If life were extinguished, there would, of course, be no problem. It is more profitable, therefore, to discuss the latter possibility, even though the former is eminently probable; perhaps the most probable.

The technico-military circumstances make world hegemony fully possible. The improvements in transport and communication and the increasing preponderance of organized armed forces over the unarmed population (the facility of suppression, as we called it) render the task of keeping down the population of the world fairly easy. The same factors, combined with the rising preponderance of attack over defence, increase the probability of wars ending with complete subjugation of the defeated parties.

The establishment of a world hegemony would mean the disarmament of all nations except one, which would become the ruling stratum of the new state. It is by no means certain that thereby social inequalities would be accentuated everywhere. A foreign conquest of, let us say, Egypt could hardly do so, because the stratification there is already extremely steep; so that all that could happen would be the replacement of the top personnel. A great deal would naturally depend on the ideology and structure of the conquering state. The Soviet conquest of eastern Europe produced some social levelling, and it is quite possible that a similar sequence of events might occur in the future. It should not be assumed that national independence necessarily leads to great social equality. Very often freedom from foreign rule means freedom for the indigenous ruling strata to oppress the masses. In Syria, to take one of the many examples which could be given, the interests of the peasants seem to have been much better protected under the French rule than they are today. Nevertheless, the establishment of a world hegemony would preclude any general levelling of social inequalities and strengthen the oligarchic tendencies. The foreign rulers, propped by their troops, would be much less dependent on the goodwill of the masses than are the indigenous governments, which have to enlist their co-operation in the task of defence. And as there would be no threat from outside, there would be no motive for extending military participation. One of the most powerful levelling factors operating in the contemporary world would thus be removed.

It may be imagined that the dominant nation might just establish its troops at strategic points, disarm other nations and leave them to rule themselves; and that they might pursue democratic and egalitarian policies. But such a situation is unlikely to arise because such domination in splendid isolation would be insecure. If left to themselves, other nations might do things which might endanger the hegemony. They might outstrip the dominating nation in industries of potential military value; they might instil into their youth the spirit of revolt and revenge; or they might even manufacture weapons secretly. A successful subjugation requires an efficient network of spies and other agents, which cannot be built without the collaboration of some of the natives. And in the conditions of a complex civilization it would not be enough to have a few hired spies. A successful domination needs a party of collaborators. Recent events prove this abundantly. The Nazis found it much less troublesome to maintain their rule in countries where they had such a party. And where, as in Poland, they outraged the people by their brutality and insolence so much that they could find only very few indigenous allies, they had the greatest difficulty in combating the underground organizations. The superiority of the Soviet method lies in its determination, unhampered by a racialist ideology, to find indigenous collaborators. The example of Poland is a striking illustration of this superiority: the Communists succeeded in stamping out all organized resistance—the thing which the Nazis could not accomplish, in spite of much greater expenditure of energy and· more indiscriminate use of terror.

But . . . why could not the victorious nation base its rule in the dominated countries on the support of democratic and egalitarian parties? The answer is that such support would be extremely unreliable. It is very profitable to seek the support of discontented masses in the countries one attempts to subjugate. Indeed, many spectacular conquests were due precisely to such support: the Ottoman conquest of the Balkans and the Greek islands belonging to Venice, Mithridates' conquest of Roman Asia, are the prime examples; also to a lesser extent the Arab conquest of the Byzantine lands and the Manchu conquest of China. The men of the Kremlin have not invented this trick, though they brought its technique to unprecedented perfection, and practice it with unheard of assiduity. But there is no single example of a domination being assured by the support of the majority. All the enduring empires sought and won the loyalty of a minority among the conquered, upon whom they bestowed privileges; perhaps the most successful in this respect was the policy of the Romans. There are very good reasons for this.

In the first place, it is very difficult to give something to everybody, or even to the majority; much more difficult than to bestow bounties upon the few. Even if all the rich are despoiled, the share of the poor in the booty must be small because they are everywhere much more numerous. In societies where the biological powers of procreation are utilized to the full, it is even impossible to feed everybody. In any case, it is most unlikely that the majority would become so enamoured of the foreign rulers as to constitute the prop of their rule because contentment, once bodily needs are satisfied, is primarily a matter of relative position. We quickly get accustomed to better things and take them for granted, and we estimate our well-being according to whether we are better or worse off than others. The desire to defend the existing order results from vested interests, that is to say, from the consciousness of having privileges to defend. The masses may defend their country ferociously, particularly if threatened with degradation by the conquerors. But they can hardly be expected to support a foreign rule when there is no immediate danger of another conqueror bringing even harsher domination. And if they were able to elect their government, they would surely vote for a party which was against this rule, and which blamed the foreigners for all their ills, be it even without the slightest justification. In any case, parliamentary democracy implies a periodic reshuffle of rulers because of the impossibility of satisfying the majority for ever. And that would mean that the foreign rulers could not choose the executors of their orders: a state of affairs which would unavoidably weaken their rule.

The party of collaborators, then, must be a small minority party, composed of men who are conscious that they owe their positions to the dominating power, and, preferably, of men drawn from the very bottom of the social ladder, and grateful for being freed from grinding poverty or even jail. The policy of the Kremlin, though it seeks popular support outside the orbit of its dominance, follows precisely this line. Foreign rulers may sometimes find support among the pre-existing privileged strata too, if these feel themselves threatened by social revolution. Such was the case of the Romans in Asia and, to take a less marked but more modern example, of the Nazis in France. But whichever variant is realized, the generalization of universal validity is that it is very profitable to seek the support of the masses in the countries one seeks to dominate, but once this domination is established, it must be based on the support of a minority; except when a newly established democratic regime is threatened with immediate overthrow from outside, as it was in some Greek cities during the Peloponnesian War. The 'democratization' of Japan under the American rule can hardly be cited as an example disproving this generalization; first, because it is more apparent than real; and secondly, because the American domination is not intended to be permanent.[1] The conclusion, therefore, inevitably emerges that the erection of a world hegemony would spell the reversion of the trends towards democracy and social equality, which have been at work in many countries during the last century . . . at least in the dominated countries.

But what about the extent of governmental regulation? How would that be affected? I have shown previously that totalitarianism is to a very large extent the consequence of polemity under circumstances calling for high M.P.R.; under conditions, that is to say, where the might of the state can be enhanced by the harnessing of the whole population. This factor would naturally

[1] Written in 1950.

cease to operate because there would be nobody to prepare war against. Nevertheless, it would be rash to predict on this basis that the world state would approach the laissez-faire type. We know that totalitarianism can arise on other than military grounds. A most notable example of such development was the Old Kingdom of Ancient Egypt. There, in spite of almost complete absence of external danger, the whole population was regimented and harnessed to the task of glorifying the Pharaos by erecting the Pyramids. It is equally possible that some absolute ruler of the future world state might conceive the desire to secure for himself immortal glory by covering the earth with even more gigantic monuments. Also, the mere desire among the officials to supervise and to boss might produce the extension of government control, a considerable amount of which would in any case be necessary because of the complexity of the economy. On the other hand, the natural indolence of most men, the universal inclination of people in 'soft' jobs to take things easy, would tend to limit the activities of officials to the unavoidable minimum. Thus might arise a society resembling the Korean Kingdom before it was 'opened up' by the Europeans and the Japanese, where even the guards at the palace of the nominally despotic king were allowed to doze off while on duty.

Much would depend, naturally, on who carried out the 'unifying' of the world. If it were done by a state wedded to a totalitarian ideology, the world state would be more likely to be totalitarian than in the opposite case. But even if a world hegemony were to be established by the Soviets, the outcome in the long run would not necessarily be a totalitarian world state. We cannot assume that what happened in newly conquered eastern Europe would happen everywhere, because, first, the difficulty of supervision grows with the bulk of dominated lands; and secondly, one of the aims of the sovietization of eastern Europe is to weld it together with the old Soviet lands into one dependable instrument of further conquests, whereas if the whole globe were subjugated, this motive would disappear. Moreover, there is no reason to suppose that the Soviet élite are immune from the disease to which all élites succumb—indolence produced by sinecures and an easy life. It must be remembered that the present élite of Russia consists of *arrivés*, of men dizzy with newly acquired power and delighting in its exercise—men who still remember the abject circumstances of their early lives; some of them may even be able to recall the feel of a bailiff's whip. But it remains to be seen what their grandsons will be like; for they will tend to take their power, riches and honours for granted, and will perhaps have acquired the art, the lack of which drives their grandfathers to work furiously, of leading idle but pleasant and interesting lives.

Concluding, we may say that, though totalitarianism would inevitably prevail in warring states, in a world state it is possible but not inevitable.

I have dealt so far mainly with the conquered lands. But what about the inner structure of the hegemonic state? What happened to it would naturally depend on what it was at the outset. Let us suppose that a state like the U.S.A., where liberalism and democracy are deeply rooted in the tradition, establishes a world hegemony. Would it be likely to continue to adhere to these traditions? It does not seem so.

Without denying the power of persistence of traditions, we must admit that they cannot resist indefinitely the pressure of the actual distribution of power. Even in the case of the Romans, one of the most conservative and traditionalist of peoples, foreign conquests destroyed the internal constitution. During the apogee of the republic military leaders were under the strict control of the Senate. But when they became almost independent rulers of extensive provinces, commanding, not now a citizen militia, temporarily enlisted and devoted to Rome's ancient traditions, but professional soldiers, personally loyal to their generals, they were able to impose their will upon the Senate.

The same would, no doubt, happen if the U.S.A. conquered the world. Its generals would be ruling despotically many countries, with the aid of professional soldiers, as these are fittest for police duties. And we can by no means be sure that the civil government would be able to control them better than it controls MacArthur today.[1] On the contrary, it seems that it would be even less able

---

[1] Since this was written MacArthur has been dismissed and the President asserted his authority. Nevertheless, the whole affair brought into sharper relief the danger of a proconsul who has found the allies within the civil government. Nevertheless, there is one very important difference between the Roman and American republics: there is no large destitute and desperate proletariat in America. Roman dictatorship (in our sense of this word) arose out of the acute struggle between the rich and the poor. The proletarians, ground down by the rich, flocked under the banners of Marius, Caesar and other commanders who were their only hope.

to control them because there would be more of them and they would need to have larger armies. The number of troops stationed in Japan is certainly no indication as to the size of the garrisons which would be necessary. The American occupation of Japan is made easy by its avowedly temporary nature and the hope of the Japanese that if they are obsequious to the Americans, they will get rid of them sooner; and also by the considerable economic help which Japan receives from its masters, which could not be given to the whole world. The permanent subjugation of the whole world would require, not only absolutely but even proportionately, far larger forces. Even if it did not—even if the whole world could be dominated by controlling a few sites for launching atomic rockets—the situation would be even less propitious for the survival of parliamentary government: the same threat could be employed against Washington as against Moscow or Bangkok. In such a situation, moreover, the secrecy which would have to envelope the army would provide an ideal ground for conspiracy. The country would be completely at the mercy of the generals, unless (what would lead to much the same results) its civilian rulers militarized themselves, like the general secretary of the Russian Communist Party who became a marshal. I must emphasize again that there are no reasons to think that mere weight of tradition could prevent the installation of an American replica of European dictators. American democracy, the essential feature of which was the subdued role of military men, was shaped during the time when every man had his gun and the army was small. Until the present time no general could be certain that, if he attempted a coup d'etat, even with the support of the whole army, he could overcome the resistance of the civilian population.

Another very important consideration is that these viceroys and their soldiers would be ruling tremendous populations despotically and by the right of conquest. The mentality engendered through the exercise of such power could hardly make them meek and amenable to control by civilian, particularly elective, bodies: they would be accustomed to disregard majority opinions.

The chances of survival of democracy and liberalism would, then, be very slim, even assuming that the U.S.A. would emerge from victory in the struggle for world hegemony with its constitution unaltered. But this assumption is surely false. As we saw before, protracted warfare would inevitably push all the contestants towards totalitarian monocracy. So that even if the constitution remained nominally the same, the actual distribution of power would undermine its viability. Taking this into account, we are forced to conclude that liberal democracy is even less likely to survive than would follow from the preceding considerations.

A totalitarian state, like Soviet Russia, would probably undergo fewer modifications after attaining world hegemony because it is better suited to total war and to global domination. It is probable, however, that it would become less totalitarian as the need to maximize military strength would disappear. Any levelling tendencies likely to persist or to develop during the era of warring states would probably disappear too, because the military participation of the masses, and therefore their loyalty, would lose its value. Social inequalities would probably be accentuated and oligarchic tendencies strengthened. These tendencies might also gain ground at the expense of monocracy because the unity of command would become less essential. In consequence interstratic mobility would probably diminish. Another reason why stratification might become steeper is that animosities between strata could not be attenuated by common opposition against outsiders.

It is generally taken for granted that the existence of a world state would automatically assure peace. This assumption is quite unjustified because civil wars might replace the inter-state wars, as happened in the Roman Empire. I have already explained that violence can be eliminated from politics only if there is a general agreement about the legitimate exercise and devolution of authority. And it is by no means certain that such agreement would exist in a state built upon conquest; particularly as the victorious nation would, in consequence of its victory, undergo profound transformations, which could not fail to disturb its ethos. That is exactly what happened in Rome. These complications might be particularly grave if the victorious state was originally democratic and liberal.

Another condition of the elimination of violence (of abiataxy, as I proposed to call it) is the adjustment of the population to resources. It is possible to keep down the toiling masses, provided the ruling stratum is united. But if the ruling stratum is prolific, and there is constant struggle for positions and exemption from degradation, violence must ensue. Probably, however, the rulers of a world state would not oppose birth control so strongly as the rulers of warring states because their power would not depend on the abundant supply of cannon-fodder. They might even actively encourage it, if they realized that in this way they might eliminate explosive forces. Their rule, therefore, might prove more stable than that of any rulers in the past. The masses would be without influence on politics and incumbent honours, but their bodily needs would be amply provided for.

The last alternative is the establishment of a World Federation. This is undoubtedly possible, though perhaps less probable than the conquest of the world by one state. True, the federations or unions of the past were created against somebody, and there would be nobody to unite against in a World Federation. But perhaps the prospect of total extinction as the penalty for continued warfare might prove at least a partial substitute for a common enemy. There is no absolute necessity for states to live in the expectation of war: Canada does not fear invasion from U.S.A., nor is Belgium afraid of France. This kind of relation might perhaps spread elsewhere . . . but only if certain conditions are fulfilled.

Before I proceed to expound what these conditions are, I must emphasize that it is completely wrong to imagine that contemporary humanity is more cruel or warlike than its forbears. The state of Europe during the nineteenth century cannot serve as a standard of normality because it was altogether unique.[1] On the contrary, the outstanding characteristic of contemporary humanity is the widespread admission that war is evil. The growing popularity of the trick of pretending to defend peace while committing aggression is proof of that. Naturally, with the progress of mech-

---

[1] Historical myopia can lead to all kinds of baseless ideas. Thus, for instance, J. U. Nef (see *War and Human Progress*, London 1950) puts forth generalizations about war solely on the basis of his knowledge of the most recent history of the westernmost tip of the Eurasian continent. No wonder that he arrives at a completely wrong notion (which, as any other banal error, is widely accepted) that the warlikeness of contemporary states is something unprecedented. Equally baselessly he attributes the recent increase in bellicosity to the weakening of the hold of the traditional religions. In reality, never were the ruling strata of the European states so irreligious as in the eighteenth century which he extols for its restraint in warfare. From the point of view of the history of mankind, we cannot regard the abatement of warfare which lasted in Europe for two centuries as anything but a blithe respite. Now we have reverted to the more usual state of affairs.

anization, mechanical replaced manual killing, but the latter method was quite effective. In proportion to the number of participants, mechanized wars have so far been definitely less bloody than many wars of old. The calculations of the incidence of war presented by Sorokin in his Social and Cultural Dynamics are completely misleading. They are vitiated by the fact that innumerable small wars which were waged in the Middle Ages and Antiquity are not mentioned in history books; we can be sure that most of them have never been recorded. The wars of primitive peoples often result in wholesale massacres. We should not accept in this matter the opinion of modern ethnographers, derived from the study of tribes already subdued by European arms. Descriptions of older travellers are much more valuable in this respect.[1] Nor can it rightly be said that it is the weakening of traditional religions which caused the recent wars. These religions have existed for thousands of years and have never succeeded in attenuating, let alone abolishing, wars. There is not the slightest reason to think that they can do so in future.

The most important drives towards war stem from the lust for power on the part of the leaders, and from the desire for wealth on the part of their followers. The danger of the first can be somewhat eliminated by representative government, even if it is only partial, because under such a system it is impossible for the rulers to disregard altogether the wishes of the masses and to use them as mere tools for quenching their thirst for power and glory. But the mere elimination of this possibility is insufficient to assure peace. It is evident in the light of what I said earlier that even a direct democracy can foster peace only if coupled with prosperity. The starving or declassed are ready for anything and will gladly surrender their political rights and liberties to a leader who will promise them booty. A real World Federation, therefore, can come into being only if there is universal prosperity. To bring that about is an enormous task, and impossible to achieve—I do not need to repeat the arguments—without the general acceptance of birth control. It would not be easy to secure this acceptance, because the practice of birth control runs counter to so many ingrained traditions: the injunction to multiply is one of the commands of many religions. The Catholic Church prohibits

---

[1] The best discussion of this point can be found in R. S. Steinmetz, *De Vredelievendheid der Laagste Volkstammen*, Mensch en Maatschappij, 1931.

this practice on the ground that it leads to undue indulgence; an understandable point of view on the part of celibate priests. Governments, moreover, are always anxious to have abundant cannon-fodder. This is not merely the result of their wickedness (by humanitarian standards). As already explained by Malthus, under the circumstances of constant warfare, generated by the pressure of the population, it is very desirable for any particular nation or tribe to be as numerous as possible; because being strong and victorious it can be prosperous too, while if it becomes weak, it may be exterminated altogether. Thus there arises the vicious circle of a demographic 'armament race' which is very difficult to break.

Another difficulty stems from the fact that birth control is not likely to be adopted spontaneously by people living on subsistence level. On the other hand, they cannot be lifted out of their poverty so long as their fertility equals fecundity (i.e. the biological maximum), because the accretion of wealth is eaten up by the increase in numbers. The fact that in western Europe the growth of wealth outstripped the growth of population should not mislead us into imagining that this is a normal process. The case of western Europe (and that of North America too) was unprecedented and will probably remain unparalleled. Possessing the monopoly of industrial technique, Europe could subjugate enormous areas overseas, and convert them into sources of raw materials and dumping grounds for its surplus population. Demonstrably, such a position cannot be attained by all countries at the same time, unless Mars or Venus are colonized. However, it is wrong to imagine that the problem would be solved if we could increase the world's food production faster than the present rate of growth of the population. Unless the birth rate dropped, every improvement in welfare would accelerate the growth of the population. Roughly speaking, if mortality in India were reduced to the western European level, its population would double itself about every twenty years. No inventions have been made so far which would entitle us to hope that the food production of India could be increased sixty-four times within a century (assuming that doubling the production of food per head would be enough for abolishing poverty). It is evident that the spread of birth control must precede any substantial rise of the standard of living in over-populated countries. This means that if that rise is to occur, the organized propaganda of birth control is required. But here again we stumble upon another difficulty. This sort of propaganda is not likely to help anyone to get into office; it is much more profitable to blame the government for all the ills. But perhaps governments might be persuaded that by conducting such propaganda they might present explosions internal or external.

It follows that any help for 'underdeveloped' areas, aiming at reducing the breeding-grounds of war and revolutions, can only achieve its goal if it brings about the acceptance of birth control. It should be granted only to those governments which take active steps in this direction. Otherwise, it is absolutely futile, or even worse: it may be helping future assailants to industrialize themselves. Moreover, a rapid industrialization dislocates social structure, weakens traditions and thus stimulates revolutionary movements, particularly if taking place in conditions of poverty. The money spent on perfecting an infallible and comfortable contraceptive appliance would undoubtedly do more to assure peace than much greater sums spent on promoting international understanding by exchange visits, etc. And one of the first items in the aid for overpopulated countries should be adequate supplies of such appliances. It is true, of course, that low fertility combined with low mortality means the suspension of natural selection, and is therefore conducive to degeneration. But natural selection has already been distorted by the progress of civilization, and modern wars are decisively dysgenic. Natural selection cannot be restored unless we go back to a completely primitive life. But we can hope that degeneration can be prevented by eugenic measures.

Concerning international understanding, I must remark that the idea that the mere multiplication of contacts must produce more cordial relations is absurd. Very often the opposite is true. People certainly do not hate each other because they do not know each other. On the contrary, the most violent hatreds exist between groups and individuals in close contact. So, though international visiting may be desirable for other reasons, it surely does not in itself promote peace.

The abolition of a tyrannical government can be only temporary if widespread poverty is not eliminated simultaneously. A tyranny can be destroyed by force of arms; it can be imposed in the same way too. But a liberal democratic government cannot be maintained by foreign armed forces; and without prosperity it is unlikely to survive by its own strength, because people will always follow him who will promise salvation from starvation through violence. On the other hand, a tyrannical and aggressive government may constitute the chief obstacle in the way of prosperity. For military reasons, it may stimulate the growth of population far beyond the economic optimum; it may direct a great portion of wealth into armaments production; it may, by its extortions, throttle the incentives to economic progress; it may maintain in power a corrupt and incompetent clique, gnawing away the foundations of economic life; and it may itself create the poverty, population pressure and resentment which will drive it further along the road of aggression and oppression. Such a government must be abolished before anything can be done about the prosperity of a country.

When one realizes how much poverty and tyranny there is in the world, one is inclined to become despondent about the prospects of a stable World Federation. Undoubtedly, World Hegemony is more probable. Nevertheless, a chance still remains that if the states which are liberal, democratic and prosperous, and at the same time powerful, make a determined, and above all skilful, effort, a stable World Federation may materialize. The amount of knowledge, diplomatic skill, far-sighted planning, patience and perseverance required for the success of such an enterprise is truly enormous. It might perhaps be available in a country ruled by a Platonic philosopher-king, but it is perhaps too much to expect from politicians, some of whom are unscrupulous and obscurantist, and chiefly preoccupied with catching votes by playing on the ignorance and blind hate which abound in any country. I shall not venture, therefore, to predict whether such an enterprise could succeed. Nor shall I offer any advice to those who wish to embark upon it; too much depends on particular unforeseeable circumstances. In any case, a useful discussion of this problem would require much more space than can be spared here. But I must say that it appears to me wrong to envisage the future solely as a contest between the U.S.A. and Russia. The whole constellation may easily change in a couple of decades. Western Europe, if united, may appear on the scene as a considerable force. China will undoubtedly emerge as a great power and, driven by the land hunger of its swarming masses, may yet become the most dangerous enemy of Russia; though that will depend on whether the diplomacy of the Occidentals proves wise and skilful enough to sow the seeds of discord among the Communist allies. Marginally, I must remark that the diplomacy of the parliamentary democratic states is incomparably more short-sighted and clumsy than that of a Metternich or a Richelieu. The community of faith is no guarantee of friendship; innumerable historical examples prove that amply. The Catholic kings of France, who persecuted Protestants in their domains, allied themselves with German and Swedish Protestants and the Turks against the Catholic Habsburgs, who in their turn found allies in the Persians, co-religionists of the Turks. The fact that the Chinese hate Europeans and Americans—and it would be surprising if they did not—proves nothing.[1] India may also become a great power and perhaps a centre of imperialist expansion. Japan, Indonesia, South America, and in the still further future, Africa may make the constellation of power even more complicated. Sooner or later they will all have atomic rockets.

But supposing a World Federation does come into being, what sort of society would then be possible? By definition, so to speak, it would be a prosperous society; otherwise it could not exist. And it would have to be free from violence. It is more difficult to guess the features of its stratification; particularly, as considerable regional variations could coexist. In some countries hierocratic or technocratic regimes might arise, while elsewhere egalitarianism

---

[1] This was written before the Korean events, but I have found no reason to change my opinion since then. These events brought to light better than ever perhaps the inevitable clumsiness of the diplomacy of parliamentary governments. In order to envisage a rift between the Russian and the Chinese governments, it is not necessary to imagine that either of them must abandon its creed. Despotic states are much less able to co-operate on equal terms than states whose rulers are accustomed to compromise.

It is impossible to guess whether such a rift will develop or not. This depends primarily on how much the Communist partners will fear one another, and how much they will hope to gain by a joint assault on others; and also on whether one of them will be able to subjugate the other. If Western countries wish to preserve their independence, their governments should do their utmost to damp these hopes of gain, and to persuade the partners that they have more to fear from each other than from the West—which is, indeed, quite true, particularly in China's case. If the two continue to stick together, the danger is that western Europe will be overwhelmed and the American continent besieged. In India, parliamentarianism has feet of clay; elsewhere in Asia it has no feet at all. Therefore, should the cleavage in a future war run on purely ideological lines, the Atlantic Union has a rather small chance of victory. But the discord among the dictators saved western Europe and America in the last war; it may do so again.

might prevail. Nor can the possibility be excluded of the supra-national police force, necessary for the enforcement of peace, seizing power. In such case the result might be the same as if a world hegemony had been established. Perhaps in either case the evolution would converge upon a form of society like that described in Aldous Huxley's *Brave New World*. One thing seems fairly certain, and that is that this world would be less dynamic than ours. It appears to me that far from living on the brink of new spectacular advances in technology we are living in an era when the rapid progress in this field is drawing to a close. It might continue in conditions of inter-state strife, so long as superior technology constitutes the most formidable weapon. But with the inauguration of the era of universal peace this stimulus will disappear. And, as I shall show in a book on the social conditions of technical progress which I am preparing, many of the forces which drove the Occidental civilization along the path of technical progress are already dwindling.[1] I cannot enter here into the discussion of structural factors, but even without this certain things are evident. In the first place, the factor which always prevented the attainment of equilibrium between population and resources—the natural tendency of the former to grow—would disappear. Secondly, the destruction of societies which did not keep up with the pace of technical progress elsewhere would cease with the abolition of war; whilst the World Authority would need to see to it that no country outpaced others in technology so much as to become a menace to them. Research in all fields connected with armaments, and that means a very great portion of the natural sciences, would certainly need to be controlled. And the control of research by bureaucrats and parliamentarians would certainly lead to its arrest. Thirdly, an all-embracing political organization would unavoidably produce a certain amount of standardization of culture. The decrease in international animosities and the widening of the freedom of travel would have this effect too. But on the other hand, the equalization of cultural prestige, consequent upon the disappearance of one-sided preponderance, would favour regional distinctiveness. Nevertheless, differences in outlook on life could not become too profound because the mere existence

---

[1] I am thinking in terms of centuries. It is possible, moreover, that philosophy and sciences which can neither endanger social structure nor provide weapons, might continue to flourish.

of common political organization requires the universal acceptance of certain norms of political behaviour and the elimination of traditions and methods of bringing up children which foster aggressiveness. And uniformity in this sphere involves uniformity in many others. Now, as intellectual progress is stimulated by contact with differing cultures, and as the elaboration of variety requires a certain measure of isolation, such standardization might well damp mental alertness. That does not mean, of course, that there would be no change whatsoever. That would be impossible because non-human. But the tempo of the change might be nearer that of Ancient Egypt or China than of the world of today. Whether that would be better or worse is a matter of ethical evaluation which cannot be adjudged from the standpoint of sociology alone. But there is no reason to think that under such conditions people would necessarily be less happy than in our world in trance, where they rush madly around, developing peptic ulcers, and suffering from mental conflicts in consequence of perpetually changing and incompatible beliefs and norms. Provided that poverty were eliminated, the world of the mandarins might not be unpleasant to live in.

Many of us who are very much imbued with the ideas of efficiency and progress find their philosophy and art not without charm. Or perhaps the world may come to resemble those idyllic pictures of South Sea Islands—people playing while they work, cultivating the arts, including those of love-making, cookery and conversation, singing, dancing and joking; leading pleasant, carefree lives, not knowing the meaning of struggle, suffering, or heroism. . . . Paradise regained!

Postscript to the Second Edition (1967)

I. MILITARISM, MILITOCRACY AND MILITOLATRY[1]

L IKE most sociological terms 'militarism' is used in several distinct senses which must be examined in turn.

1. 'Militarism' is sometimes taken to mean militancy or aggressive foreign policy involving the readiness to resort to war. It does not seem profitable to'adopt this usage as the words 'militancy' and 'aggressiveness' adequately describe these features.

2. In other contexts 'militarism' means preponderance of the military in the state. Such a preponderance implies a differentiation of civil and military spheres of authority, and of civil and military administrative personnel. It would be improper, therefore, to apply the term 'militarism' in this sense to situations where such differentiations are absent, as was the case in all primitive states such as, for instance, the Zulu and Ankole kingdoms in Africa, or the Polish kingdom under early Piasts, or even the enormous empire of Gengiz Khan. Whereas the differentiation of the roles of the chief priest and the war leader occurs even in small tribes, the first trace of a distinction between military and civil (to be exact—financial) spheres of authority exercised over the same subjects, appears in the Persian Empire in the fifth century B.C. Although some American Indian tribes had war chiefs as well as peace chiefs, this arrangement constituted neither a division of authority over the same persons nor an allocation of the population to permanent units; it could be best described as an alternation of types of authority.

To distinguish between this and the third interpretation of the term 'militarism' I venture to suggest that the phenomenon of preponderance of the military over the civil personnel be called 'militocracy'.

[1] Based on the entry prepared for *A Dictionary of the Social Sciences* (Tavistock Publications, London, 1964).

POSTSCRIPT TO THE SECOND EDITION

An important feature distinguishing different kinds of militocracy is the extent to which the rank and file share the privileges of those at the top of the hierarchy. Variations in this respect depend on military participation ratio. Obviously, if military participation ratio is very high the privileges of the rank and file have to be diluted to the point of non-existence. Germany under Wilhelm II and Poland under Pilsudski exemplify a variant of militocracy where political preponderance and economic favours were restricted to the officer corps. The Late Roman Empire, on the other hand, is an instance of the more inclusive variant, the distinctive feature of which is the inclusion of all soldiers in the priviledged body. The connection between militocracy and bureaucracy depends on the type of military organization.

Contrary to widespread opinion, militocracy is not necessarily accompanied by external militancy: Tokugawa Shogunate in Japan, as well as a number of Latin American military dictatorships, are examples to the point.

3. 'Militarism' can also be interpreted as connoting the extensive control by the military over social life, coupled with the subservience of the whole society to the needs of the army. This leads usually to a recasting of various aspects of social life in accordance with the pattern of military organization. It seems, however, that we might reduce the danger of ambiguity if we use the term 'militarization' instead of 'militarism' to describe this phenomenon.

Militarization can occur without militocracy, as can be seen from the examples of Britain and the U.S. during the Second World War: nevertheless it seems that this can be so only in the short run. On the other hand, many cases show—a number of Latin American dictatorships, for instance—that militocracy can endure without producing a wholesale militarization.

4. It has been proposed by some writers that by 'militarism' we should mean the pointless, or even harmful from the point of view of efficiency, addiction to drill and ceremonies, and adulation of trappings. But the tendency towards a shift of valuation from ends to means, and from content to form, is a ubiquitous social phenomenon; and 'militarism' in this sense is thus merely a manifestation of this tendency in the military field. Thus usage seems to be unprofitably restricted.

5. Sometimes the word 'militarism' is used to refer not to an institutional arrangement but to an ideology propagating military ideals. Such an ideology often accompanies militocracy (e.g. Germany under the Hohenzollerns, Japan under Tojo), yet the latter can occur without it (e.g. Cuba under Batista, where the soldiers were despised rather than admired). Furthermore, an ideology extolling the soldier and the military virtues may flourish even where the army is weak (e.g. Germany at the time of Weimar Republic). For the sake of clarity, it would be better to speak of 'militolatry' than of 'militarism' in this sense.

6. There is an interesting phenomenon which, however, it would be inappropriate to call 'militarism'; namely, the inclination to imitate military demeanour and paraphernalia in the walks of life entirely unconnected with war. The example of the Salvation Army shows that such a 'militarism' can flourish even where militarism in other senses is not prominent.

7. Sociological analysis would be facilitated by defining militarism as the compound of militancy, militarization, militocracy and what might be called militolatry, that is to say adulation of military virtues. Where all four components are present to a high degree (e.g. Japan under Tojo), we have a clear case of militarism. Different types of militarism could be distinguished in accordance with the relative strength of the components.

2. ORIGINS OF WAR[1]

All types of human activity must have emerged at some point of time, but whereas some of them (like science or machine industry) came into existence in the full light of historical records, the origins of others are enveloped in eternal obscurity. The origins of capitalism or even the art of writing constitute a subject for argument based on evidence, but only tenuous conjectures can be offered on the problem of the origins of language, religion or war.

The question of the origins of war is even more hopeless than that of the origins of the state, because we know a great deal about

---

[1] Reprinted and abridged from a contribution to a symposium *The Natural History of Aggression* published for the Institute of Biology by Academic Press, London and New York 1965.

primitive groups and tribes which are stateless in the sense of being without anything that can be called a government. Moreover, there are historical records describing a number of cases of emergence of a state. Nothing comparable exists as far as war is concerned. There are, of course, inexhaustible records narrating the origins of particular wars, but war as a pattern of activity antedates by far the art of writing, and therefore the problem of its origins in the strict sense is insoluble. At most we can consider what might be the causes of its ubiquity.

The systematic killing in which mankind indulges, is at variance with what goes on among other mammals. One of the chief reasons for this difference is the obvious circumstance that human beings use weapons; without which killing need not be a by-product of fighting, because in such a situation a victor does not have to fear the vengeance of the weaker opponent. But this is not so if weapons are used, because then he who stabs or shoots first wins, and under such circumstances it is safest to kill one's enemies. Anyway, in all fighting where weapons are used some of the participants are likely to get killed. So we are justified in saying that the prevalence of killing within our species was made possible by the acquisition of culture.

War has been blamed on human nature, and it is perfectly true that if all men were kind and wise there would be no wars. It is clear that the capacity for cruelty is required for war, and the proneness to collective follies always facilitates wars and other kinds of social conflicts. Fortunately, however, there are reasons for doubting whether war is an absolutely necessary consequence of human nature being what it is. In every warlike polity (which means in an overwhelming majority of political formations of any kind) there are elaborate social arrangements which stimulate martial ardour by playing upon vanity, fear of contempt, sexual desire, filial and fraternal attachment, loyalty to the group and other sentiments. It seems reasonable to suppose that if there was an innate propensity to war-making, such a stimulation would be unnecessary. If human beings were in fact endowed with an innate proclivity for war, it would not be necessary to indoctrinate them with warlike virtues; and the mere fact that in so many societies past and present so much time has been devoted to such an indoctrination proves that there is no instinct for war.

Another important point is that in many nations for decades and

187

even centuries only a very small minority of men took part in wars, so that the statistics of wars give a very exaggerated picture of the prevalence of bellicosity. For example: during the nearly three centuries between the death of Cromwell and 1914, Britain waged dozens of wars (or even hundreds if we include the colonial expeditions) but the soldiers who fought in battles constituted much less than one per cent of the population and even they spent much more time in the barracks than on the battlefields. Nor is there much evidence that the majority thus condemned to a peaceful existence felt an irresistible desire to participate in wars, although it is true that many wars began with an outburst of collective enthusiasm.

It must not be forgotten how often direct compulsion had to be used not only in recruiting soldiers but also to make them fight. One of the masters of the craft—Frederick II of Prussia—enunciated the principle that a soldier must fear his officer more than the enemy; and according to Trotsky a soldier must be faced with the choice between a probable death if he advances and a certain death if he retreats.

If men had an innate propensity towards war, similar to their desire for food or sexual satisfaction, then there could be no instance of numerous nations remaining at peace for more than a generation. Nor can war be regarded as an inevitable consequence of national sovereignty because there are examples of sovereign states which have waged no wars for more than a century—these are Switzerland, Sweden, Norway and Denmark. One could say that with Switzerland, surrounded as it is by much more powerful neighbours, peacefulness is a matter of necessity rather than choice, but in so far as the Scandinavian countries are concerned, it is clear that although they were too weak to attack their neighbours to the south, they could have fought among themselves, as many other small states did. There are a few other examples of peacefulness when a conquest would be very easy. The United States, for instance, could conquer Canada without much effort or fear of reprisals; nevertheless, the Canadians have no fear of such a possibility. In spite of being exceptional, these examples do show that truly peaceful co-existence is possible.

It is often claimed that the remedy against war is to institute a world government, but this view can be easily refuted. In the first place, political unification often merely means that instead of inter-state wars, civil wars take place which can be just as bad or even worse. To mention one of very many possible examples: as soon as the Romans had defeated their dangerous enemies, they started fighting among themselves, thus bringing the lands which they 'pacified' into a much worse condition than they were in when they were divided into a multitude of independent and warring states. During the last hundred years the countries which waged fewest wars—Spain, Portugal and the republics of Latin America—had the longest record of internecine strife and revolutions. For this reason we cannot assume that we could eliminate bloodshed simply by instituting a world government, because the outcome would depend on whether the sources of strife and violence would be eliminated.

Even more: it seems that there is an inverse connection between strenuous warfare and pretorianism. In Rome the civil wars began when Carthage—the last enemy who could threaten the very existence of the empire—had been destroyed.

In modern Europe the country which was most plagued by revolutions—Spain—did not take part in any major war since the times of Napoleon. What is even more telling is that it acquired this propensity only when it ceased to conquer and to send colonial expeditions. On the other hand, in Russia, which was nearly always at war and conquering, the army remained remarkably obedient. The Japanese, who remained confined to their islands nearly throughout their history by the overwhelming might of the Chinese empire, have an unrivalled record of civil wars. Latin America, where very few wars were fought, experienced more military revolts during the last century and a half than the rest of the world put together. The inverse connection between outward militancy and proneness to revolts, which the foregoing evidence suggests, is explained by the fact that external and civil wars are alternative releases of the pressure of population on resources.

It must be remembered that wars against rebels constituted, next to external wars, the chief occupation of governments throughout history. On the other hand, the only really peaceful area of the world—Scandinavia—has no supreme authority, the real cause of its peacefulness being that it is free from poverty and despotism. The same is true about internal peace: only countries where there is neither poverty nor despotism do not suffer from internal violence.

In conditions of misery, life, whether one's own or somebody else's, is not valued, and this facilitates greatly warlike propaganda. In an industrial society unemployment not only brings poverty but also breaks up social bonds and creates a large mass of uprooted men, whose frustrated desire for a place in society may lead them to favour measures of mass regimentation. Moreover, when there is not enough to satisfy the elementary needs of the population, the struggle for the good things of life becomes so bitter that democratic government, which always requires self-restraint and tolerance, becomes impossible, and despotism remains the only kind of government that can function at all. But absolute power creates the danger that a despot may push his country into war for the sake of satisfying his craving for power and glory.

The rulers who embark upon aggression on their own initiative are prompted chiefly by their desire for more power and glory— by the wish to be above their opposite numbers—and directing a war can be fun for a callous despot. Louis XIV—to mention one of innumerable possible examples—used to start a war whenever he was bored, without, of course, exposing himself to any dangers or privations. Unlike contemporary despots, he was quite frank about it. It follows that one condition of abolition of war is the elimination of situations which permit rulers to amuse themselves in this way at the expense of frightful suffering of their subjects: in other words, elimination of despotism. It is quite clear, unfortunately, that the existing despotic states cannot be transformed from outside, and our only hope is that they might gradually evolve into more humane forms of government.

As far as ordinary people are concerned, who have to endure all the sufferings, their most important motives in supporting aggression are: (1) collective frenzy or (2) simple obedience combined with the herd spirit, or (3) a sense of desperate frustration which makes them covet other people's goods and welcome all adventures. Usually these factors are intertwined. The mass movements assume forms of collective mania chiefly in response to extreme frustration of elementary needs, including the need to have a secure place in the social order. Such frustration is most commonly the consequence of poverty, or at least of impoverishment in relation to customary standards. If that is so, we need not be surprised that war was a permanent and universal institution, for

poverty was, everywhere—and still is in most parts of the world—
a permanent condition of the great majority. Only in very recent
times, and only in the few fortunate countries bordering the north
Atlantic, has grinding poverty become rare.

The remedies of signing treaties of eternal peace, convening con-
gresses and preaching condemnation of wars have been tried innumerable times and without much effect. They may be needed,
but in themselves are clearly insufficient. Elimination of poverty
has not yet been tried except in very restricted areas, where it had,
in fact, the result of instilling into people a relatively pacific
disposition.

Given the propensities of human nature, the tendency of the
population to grow beyond the resources has ensured the ubiquity
of wars, although not every single instance of war had this factor
as an immediate cause.[1] Wars might cease to be a permanent
feature of social life only after the restoration of the demographic
balance whose disappearance at an early stage of cultural development made them inevitable.

3. PREDATION VERSUS CONSTRUCTION[2]

Parasitism exists in all human societies: everywhere there are
people who succeed in obtaining a large share of wealth without
contributing in any way towards its production. There are, however, differences of degree which are of decisive importance: in
some societies it is a residual phenomenon whereas in others it
pervades the whole social fabric. Generally speaking, parasitism
constitutes the most powerful brake on economic progress by
destroying the link between the effort and the reward.[3]

[1] An analogy might make this point clearer: the fact that everybody who survives to
a certain age enters a phase of senile decay insures that everybody eventually dies. The
statement that senescence is the cause of men being mortal remains true even though
only a minority (perhaps very small) die simply of old age. In the same way, the
statement that the tendency of the population to grow faster than the resources makes
wars inevitable is in no way invalidated by finding that many wars have occurred
without this factor being in operation. Most of the criticisms of the demographic theory
of conflict are based on the elementary logical error of imagining that a proposition
'A implies B' is invalidated if we find that B has occurred without A.

[2] The following exposition is based on the arguments developed in *The Uses (Elements)
of Comparative Sociology* and *Parasitism and Subversion*.

[3] This point is treated in greater detail in *Elements of Comparative Sociology*, Chapters
11, 15 and 16.

Once a society is pervaded by parasitic exploitation, no one has other choice than to skin or be skinned. He may combine the two roles in varying measure but he cannot avoid them: he cannot follow Candide's example and till his garden, relying on hard work for his well-being, because he will not be left alone: the wielders of power will pounce upon him and seize the fruits and tools of his labour.

The greater and more general the poverty, the more important are the stakes of political contests. In an opulent country a man who is thrown out of office can usually find decent employment elsewhere, but in a poor country, full of paupers and genteel semi-paupers, a loss of office usually means ruin for anybody who has no private weath; and even the people of the latter category are safe only in so far as the rights of property are respected. In consequence, the fight for offices becomes a matter of life and death for all the minor figures who cannot place large funds in foreign banks, and assumes the form of a struggle for existence fought with every available means and without regard for law or convention.

Under such circumstances the politics tends to oscillate between despotism and anarchy because a constitutional parliamentary government cannot function without a readiness to compromise and to observe the rules of the game. This applies even to oligarchy, but as far as democracy is concerned it can be stated as a general rule, to which no exception has yet been found, that it can function only in a society which is fairly prosperous, and in which privileges obtained through political influence are not indispensable for making a decent living. Despotism, however, cannot ensure peace for periods exceeding the length of one reign unless it is based on an undisputed rule of inheritance.

Economic progress requires peace and order. Planning, saving and investment become unprofitable if nothing is secure. Strife impoverishes and may ruin a country, if it is sufficiently serious. On the other hand, it is equally clear that generalized poverty, and especially sudden impoverishment, generate strife and disorder; so that a vicious circle maintains a country in a pitiful condition. However, the relation between economic conditions and ruinous strife is complicated by the influence of the factor of the strength of political structure: a gust of economic adversity may break a weak political structure whilst a strong structure may withstand it.

The causal relationship between parasitism and poverty is not unilateral but circular.[1] The principle of circular causation between parasitism and poverty arises from a combination of Charles Comte's principle that parasitism causes poverty, with the principle that in a fairly complex society poverty fosters parasitism. The qualification concerning complexity is necessary because poverty has no such effect in simple unstratified tribes: parasitism can flourish only in societies in which there are differentiated groups, some of which can exploit others. Instead of 'causes' the word 'fosters' is used in the second component, because the effects of poverty upon parasitism are less immediate than the effects of parasitism upon poverty, and can be temporarily counterbalanced by such factors as ideological fervour or the power of an austere despot: in Russia under Lenin there was more poverty than under the Tsars, but less parasitism.

The assertion that poverty produces parasitism does not imply that parasitism cannot grow in its absence. Actually, bureaucratic parasitism is growing in the wealthiest countries and, although it is probable that eventually it may produce an arrest of technical progress, at present it is even stimulating it by preventing crises of over-production, such as the one which nearly destroyed capitalism in the thirties. In order to clarify the issue we must make a distinction between a relatively benign parasitism in opulent societies, where the parasitic existence of a fairly large number of people does not necessitate severe deprivation of the rest, and more malign forms which occur in poor societies, where the unproductive and comfortable existence of a minority can be secured only by ruthless exploitation of the majority. Putting it into medical parlance, we can say that tolerance of parasitism (that is to say, immunity to deleterious effects thereof) depends upon wealth. The medical analogy is exact: a dose of bacteria which may be fatal to a debilitated body may scarcely affect a strong one, but nearly every illness increases the vulnerability to other diseases.

Granted that poverty in complex societies is in the long run invariably accompanied by parasitic exploitation, it may be argued that this correspondence is accounted for by the principle that parasitism breeds poverty. There are, however, examples of how impoverishment caused by extrinsic factors stimulated parasitism, which then produced further impoverishment. This was notably

[1] Exemplifications of this principle can be found in *Parasitism and Subversion* and in *Neocolonialism and Kleptocracy*.

the case with Italy at the close of the Renaissance, when a shift in the trade routes reduced commercial opportunities and stimulated the conversion of entrepreneurs into landlords and rentiers. Secondly, if the relationship between poverty and parasitism were in the nature of one-way causation, then it would be much easier to abolish them: it is because they are enclosed in a vicious circle, entangled with other vicious circles, that this is so difficult. (We can deduce from general principles of cybernetics that in a system of interdependent variables, containing stochastic elements, only those variables can be consistently maintained at a maximum or minimum which are included in positive feed-backs (i.e. vicious or virtuous circles). Thirdly, we can trace the social mechanisms through which poverty produces parasitism.

The said mechanisms exemplify the principle of the least effort: all men seek the wealth necessary for the satisfaction of their basic needs, and everywhere where wealth can be conserved there are men who amass it in order to gain more power and glory. If the easiest or quickest road to minimal prosperity as well as to riches leads through participation in activities which add to the collective wealth, then men will put their energies into socially useful occupations. If, owing to circumstances among which general poverty usually occupies a prominent place, productive activities are unrewarding, then men will concentrate on devising ways of wresting from the others such wealth as already exists. In other words, the energies which in an expanding economy will be applied to production, in a stagnant or contracting economy will be canalized into open or veiled predation. Naturally, everywhere there are people who will always opt for parasitism and predation, and others who will never do so; but the great majority can be swayed by the relative advantages of these alternatives.

The social mechanism in question offers an analogy to the mental mechanism brought to light by Freud (with whose detailed interpretation thereof we need not agree): when the basic propensities of human nature find no outlet, the mental energy turns inward to consume itself in internal conflicts.

Another mechanism of conversion operates between internal and external conflicts. It can be succinctly described as follows: external and internal conflicts represent alternative manners of predation; they constitute alternative and mutually compensatory releases of population pressure, as they are alternative methods of organizing emigration to the hereafter; finally, by displacing resentments and aggressiveness towards the outsider, an external conflict helps to smooth internal quarrels, and vice versa. Thus, social energies can be regarded as having three main outlets: construction, internal strife and external conflict. The relationship between them is such that a blockage in any of them produces an increased flow through the others, whereas a widening of one tends to drain off the flow through the other channels.

4. CONSERVATISM AND RADICALISM OF THE MILITARY[1]

The majority of European critics of the traditional social order have taken it as self-evident that military officers, by the very nature of their profession, were bound to be conservative, that is to say, always inclined to throw their weight on the side of the established social order and the privileged classes. The idea that an army could become a chief engine of social revolution would seem to them absurd. This opinion was not—and is not—without foundations, yet it cannot claim absolute validity. In the Near East we saw recently social revolution carried out by army officers. In Ancient Rome traditional aristocracy was decimated and despoiled by the soldiers on more than one occasion. This poses the question of the factors which determine whether the influence of the army will be conservative or radical, or even subversive.

Subversiveness is a rather loose and emotive term, and for this reason some terminological clarification might be useful. An established order of privileges may be altered in three possible ways. The first is a transformation which changes the whole nature of the privileges, besides producing vertical mobility on a large scale. A thorough industrialization of a rural society, or a replacement of private by state ownership of the means of production, are examples to the point. The second type of change is a process of levelling. Almost needless to say, this means that the inequalities diminish—not that they disappear altogether—for no case of complete abolition of all privileges has so far been noted. The third possibility is the replacement of the incumbents without any profound changes in either the nature of the privileges or in their span. We can safely leave out of consideration profound

---

[1] Reprinted with minor alterations and omissions from *The European Journal of Sociology*, II (1961).

transformations, for the simple reason that in no case have they been produced by an army's intervention in politics. The social reforms which ensued from the recent military revolutions in the Near East do not really amount to basic transformations. The Meiji reforms in Japan, though precipitated by a show of military helplessness, were a complex and comprehensive process and cannot be cited as a case to the point. We are left, then, with the processes of levelling of privileges and of displacement of the incumbents. By subversiveness I mean the inclination to bring about either of these modifications of the social circumstances. Our judgment on the ethical value of subversiveness as thus defined will depend on how we evaluate the established order and the alternatives.

An army may be employed against the enemies outside or against their own civilian population. Both of these modes of employment produce modifications of the social structure, but they do it in entirely different ways. A war, particularly when the very survival of the state is at stake, produces an adaptation to the requirements of bellic efficiency, whose nature depends above all on the technique of warfare. The need for efficiency may impose considerable levelling of social inequalities, as well as an increase in vertical mobility; but it may also have an opposite effect. In the cases of adaptation to bellic efficiency the impact of the army does not consist of sheer compulsion, and in the extreme cases may even be backed by the enthusiasm pervading the whole population. This mode of influence of the armed forces on society might be called consociative. The other mode of influence, the essence of which is the forcing of an unwanted social order upon the civilian population by the use or threat of violence, might be termed coercive. Almost needless to say, these distinctions specify 'pure or ideal types', and any concrete case presents an intermixture of both, though in widely differing proportions.

Where the armed forces constitute the ruling class there can be no question of their subversiveness. Neither in medieval Europe nor in the Ottoman Empire, nor in any other society of similar type, did the warriors menace the social order, in spite of their frequent rebellions. They fought among themselves for spoils and glory, but never took the side of the humble against the rich. The possibility of assault upon the social order arises only when there is a differentiation between the civilian and the military sectors.

Gaetano Mosca considered the subordination of the armed forces to the civilian authority to have been one of the most distinctive and crucial features of the European civilization. Although many military dictatorships have arisen since he wrote *Elementi di Scienza Politica*, there can be little doubt that, in comparison with what went on in other parts of the world, ever since the end of the Middle Ages the European armies were on the whole very obedient. The number of European monarchs who have been killed by their troops is negligible in comparison with what happened elsewhere in the world. In spite (or perhaps because) of its anti-militaristic ideology, China suffered a number of military rebellions on a scale unparalleled in Europe: the revolt of An Lu-Shan during the reign of the T'ang dynasty was a social cataclysm of the first magnitude. Mosca's explanation of this peculiarity of the Occident was that it was due to the integration of the armed forces in the body politic, and particularly to the integration of the officers' corps with the ruling class. Let us for the moment disregard the other ranks and examine the validity of Mosca's thesis in so far as it concerns the officers. In a way it seems paradoxical to explain somebody's subordination by the fact that he belongs to the ruling class. Moreover, everywhere and always the commanders of armies enjoyed opulence and other privileges, and in this sense they belonged to the top layer of society. The question is whether in addition to this they were united with the rest of the ruling stratum by the bonds of kinship, common outlook and habits, and of common economic interest. Mosca's explanation does appear to hold if we compare Europe with Asia. Naturally, one must be very careful when speaking of the whole of Asia, for the circumstances varied enormously in time and space, but it does seem that, unlike Europe, Asia was not free from the phenomenon which might be described as the alienation of the office-bearers. In Europe, the monarchy and the nobility found on the whole a *modus vivendi*, and created a code of rights and obligations which was fairly well observed, whereas in the Orient (with the partial exception of China) a violent struggle went on continuously between the rulers and the magnates. In the deadly struggle against the magnates, the rulers often employed slaves and mercenaries recruited from the lowest strata. These troops revolted frequently and on some occasions deposed the rulers, decimated and despoiled the nobility, and put themselves in their place. In this way, even before the Ottoman conquest the Turks replaced the Arabs as the ruling ethnic element in the Near East. By a similar process the Slave dynasty established its rule in Northern India, and the Mamluks in Egypt. Such a thing has never occurred in Europe since the times of Ancient Rome. What did occur in Rome fits perfectly Mosca's thesis—which is not surprising in view of the high standard of his classical scholarship. So long as the senatorial nobility furnished officers it was secure. Marius and his soldiers were proletarians, or at least of proletarian origins. The generals and the soldiers who exterminated the Roman nobility in the third century A.D. began their lives as sons of peasants in the outlying provinces. Here, however, it was not so much the ruler's tactics which excluded the nobles from higher military posts as their own inaptitude and preference for more pleasant ways of life. It must be remembered that ancient wars were much harder and more dangerous than the wars waged in modern Europe until 1914.

Unless an army is too weak to be able to impose its will on the civilian population, as in the United States in the last century, the integration of the officers' corps in the ruling layer appears to be a necessary condition of some measure of political neutralization; but the evidence from Latin America shows that it is not a sufficient condition, for there pretorianism flourished notwithstanding the monopolization of the higher military posts by the landowning families. This was due in the first place to the uncrystallized character of the political order, to the lack of any accepted code of behaviour on political matters, to disorderly administration and the absence of well-organized political parties, which made violence inevitable. In Chile, where the ruling class was cohesive and more orderly, the military remained at the back of the stage. The second cause of the predominance of the military in Latin American politics has been the bitter class warfare raging there ever since the days of the conquest. A very important factor here was the absence of the need for united effort in a common struggle against foreigners, which did so much to infuse a strong dose of supra-class solidarity into the European nations. It is sad to find that the absence of concern for the common good might be the result of relatively unwarlike existence. Apart from warranting the foregoing qualification, the data on Latin America do not contradict Mosca's theses: the armies did not remain in the background of politics, but they were very far from being subversive. On the contrary, they invariably intervened in the class struggle when the ruling classes were losing their grip on the masses. The exceptions to this rule also fit Mosca's thesis. The very few attempts at some levelling of social inequalities made by the armed forces were initiated or led by officers who were not closely related to the very rich; and who were, moreover, below the rank of general. 'The social identification with the urban groups where he originated was probably the fundamental cause of the junior-officer uprisings that occurred in Latin America's armies in the second quarter of the twentieth century.' [1]

The case of Egypt fits perfectly Mosca's thesis. Nasser and his colleagues did not belong to the hereditary rich, who shunned military service. The honorific as well as economic position of an army officer was low, so that the officers shared, and later canalized, popular discontent. The Iraqi revolution was headed by Kassem, a general and a man of somewhat more substantial family background than Nasser, and yet his government was and is more radical than the Egyptian in the way in which it attacked the old ruling class. However, it is well known that social determination of attitudes has only a statistical validity, and always admits individual exceptions. The social origins of the majority of Kassem's assistants seem to be even more demotic than those of Nasser's. This is even truer about the leaders of Syrian military socialism in 1966.

'Almost invariably, Latin America's popular revolutions of this century were led by the young officers. They became the sponsors of fundamental change and reform, the underminers of traditional institutions, the proponents of public-welfare measures ... These young officers thought of themselves as enlightened members of a new, modern generation. Regarding the generals as unimaginative and behind the times, they sought to bring the armed forces into more sympathetic relations with the rest of the society. They were also interested in power, which could be had by gaining popular support, by playing the role of saviours of the downtrodden masses.' [2] There are examples from other areas of the world (the young Turks, the Decembrists in Russia, Nasser's team), which show that, on the whole, the political interventions

---

[1] Edwin Lieuwen, *Arms and Politics in Latin America* (New York, Praeger, 1960).

[2] Lieuwen, *Arms and Politics*, pp. 128–9. See also *Parasitism and Subversion*.

of younger officers tend to be less on the conservative side than those of their seniors. This may be due to the natural conservatism of old age, or to greater satiety induced by having reached the summit of the hierarchy. However, there are good grounds to believe that youth as such infuses a certain rashness and extremism into political action, but does not decide whether it will be carried out for the benefit of the privileged or of the hampered. Young men are, no doubt, not only more rebellious, but also more susceptible to the appeal of impersonal ideas, than their elders, yet they can very well rebel against social equality, and the ideals which they uphold might be elitarian in the extreme. In Japan in the thirties, the young officers terrorized their seniors as well as the civilian politicians into adopting a policy of aggression outside coupled with the repression of the liberal and pro-labour elements at home. The differences in rank and age are particularly likely to foster socially subversive ideas when accompanied by differences in class affiliations, especially if demotic affiliations constitute an impediment to promotion. Apart from the cases mentioned above, this was the situation of Napoleon Bonaparte and of a considerable number of other officers of the French Army on the eve of the Great Revolution.

Integration or alienation may determine the attitudes of the officers, but for effective action they must have control over their men. There are examples of *coups d'état* which failed because the soldiers did not obey the orders, and remained loyal to the existing government. This was notably the case with the Generals' Revolt against Hitler in 1944. The second danger is that, by ordering the soldiers to disobey the established authority, the commanders may provoke a general rebellion which might turn against them as well. There were examples of this during the Wars of Independence in Latin America. Let us examine the circumstances which make either of these dangers more or less imminent.

The first factor to be considered is the strength of the links which bind the soldiers to the civilian society. Other things being equal, the weaker they are, the more absolute is the control of the officers over the men in political matters. The alienation of the soldiers from the civilian society may be the result of long-term service, or of foreign origins, or of methods of recruitment which select a special psychological type, or of stationment in distant lands or in the midst of a cultural environment with a distinctive ethos. One or more of these features characterized all military formations renowned for incursions into politics: the Roman pretorians, the mercenaries of the Islamic monarchs, the French parachutists in Algeria. In Ancient Rome military revolts began when the army acquired these characteristics. Psychologically un-selected short-term conscripts can be used for this purpose when either the whole population is ignorant or politically apathetic or when they are drawn from an unassimilated ethnic minority. The troops which Pilsudski used for destroying the constitutional government in Poland came from the eastern territories, which were not ethnically Polish; the troops which opposed him came from western Poland, where there was only a German minority which was able to evade military service. In Latin America the lowest ranks are either mercenary (as they were in Cuba) or re-cruited among illiterate peasants. In Peru, Colombia and other 'mestizo' countries these peasants have the further advantage of being Indians who have to be taught Spanish.

The alienation of the army largely determines whether it can serve as an instrument of rebellion without the danger of a chain reaction which would spread into the civilian population. The same applies to the internal struggles within the army. When in 1933 the N.C.O.s and privates of the Cuban Army, led by Sergeant Fulgencio Batista, imprisoned or killed their seniors, and replaced them as the rulers of Cuba, the civilian population stood aside: the affairs of the mercenaries were of no concern to them.

The strength of any social body depends on its solidarity, which is not fostered by extreme inequalities and the absence of upward mobility. That the latter circumstances weaken the nations has been noticed long ago. At the beginning of the last century a Polish historian, Joachim Lelewel, attributed the fall of Poland to this cause (without, of course, using these terms). It was the proof of Mosca's genius that, diverging from all current opinions, he put forward the view that the relative docility of the European armies depended on the rigidity of class divisions in their midst. He reached this conclusion by comparing them with the armies of the Roman Empire. Military commanders of modern Europe could not easily defy their rulers because, being divided from their men by an impassable economic and cultural abyss, they did not enjoy the popularity of their Roman counterparts. To put it into psychological parlance, there was no basis for projective identification. The uniting force of common hopes of enrichment was lacking too.

To the foregoing arguments a proviso must be added, that the internal horizontal chasm only hampers the intrusion into politics of the armed forces, it does not make them altogether impossible. In the same way the integration of the rank and file in civilian society makes them less suitable as an instrument of *coups d'état*, yet they can be so used if the issues do not arouse strong passions among the population. Other factors which neutralize the importance of the soldier's convictions are the discipline, the efficiency of the organization and the unanimity among their commanders.

An important question which must be considered is whether wars or preparations for them foster or impede military intrusions into politics. The chief difficulty in answering this question stems from the ways in which the impact of war is connected with other factors such as the growth and vertical unification of the armed forces, which strengthen an army's internal position irrespectively of whether a war is being waged or not.

Thus it frequently happens that when a war ends, the balance of power between the military and the civilian authorities finds itself strongly tilted in favour of the former. Moreover, the afterglow of the apotheosis of the military chiefs, which normally takes place during a war, endows them with political prestige which they could not acquire in time of peace. This is the origin of the father-figures of Hindenburg, Wellington, Eisenhower, de Gaulle. If a defeat has undermined the standing of the rulers, or if the country is infested by ex-combatant desperadoes who can find no satisfactory place in civilian life, an aftermath of war may generate disorders which automatically bring the armed forces into the arena. In these ways, then, wars stimulate military interventions in politics. On the other hand, however, 'the devil finds work for idle hands': the soldiers who have no wars to fight or prepare for will be tempted to interfere in politics. Taking a long-term view, it seems that there is an inverse connection between strenuous warfare and pretorianism.

POSTSCRIPT TO THE SECOND EDITION

5. MILITOCRACY IN AFRICA AND LATIN AMERICA[1]

Until recently one could not express a doubt about the reality of democracy in the new African states without being called a racialist and fascist. In a way this was understandable because as commonly used the word 'democracy' has long ago ceased to have any definite meaning, and has come to indicate a mere approval of the given system, whatever it might be. The debates about what democracy 'really' is in fact amount to quarrels about the rights to use the approval-eliciting verbal sign in order to influence people's attitudes. If we want to use this term for the purpose of enlightenment rather than propaganda we must specify what we mean, keeping as close as possible to the etymology, but not adhering to it completely, because 'government by the people' has never existed and in all likelihood never will: everywhere some people make decisions and others carry them out. In the following discussion, I shall mean by 'democracy' government based on the periodically expressed consent of the majority—in other words, a system of government where there are effective institutional arrangements sanctioned by laws and customs, which prevent the holders of supreme authority from continuing in office if they are actively disliked by the majority of the citizens. Elections and the plurality of parties constitute important elements of such arrangements, but they can be made ineffective by collusion, fraud, intimidation and bribery. Democracy, defined as government by consent of the majority, prevents only an ill-treatment of this majority, and is perfectly compatible with the oppression and exploitation of foreigners or internal minorities, with bellicosity, intolerance, obscurantism and many other evils. Any political system requires specific conditions for its operation, but we should not include them in the definition of the system. As a general methodological rule, definitions should be as simple (or, to use a more technical term, as unintensive) as possible in order to avoid pre-judging by definition what should be left as an open question to be decided in the light of empirical evidence. Thus, for instance, it is by no means obvious that social mobility always promotes democracy as defined above; although many people are inclined to call democratic any system where the élite are recruited from the lower

[1] The social and economic context of this phenomenon is examined in *Parasitism and Subversion* and *Neocolonialism and Kleptocracy* on which this section is based.

classes or where some levelling of the distribution of wealth or abolition of hereditary privileges has taken place. An authoritarian government may be populist in the sense of genuinely attempting to satisfy the desires of the masses, and it may (as was the case with Hitler or Peron) enjoy massive support, but talking about democratic dictatorships or totalitarian democracy deprives these terms of any recognizable meaning. Totalitarianism can be egalitarian but not democratic.

The process of creating the new African states has provided a vivid illustration of word magic. The usual precedure was as follows: a committee of lawyers would be set up who would draft a constitution containing all the democratic provisions they could think of. Then the rebellious African politicians were told that they could have independence and power if they signed a promise to follow it. Signing a small piece of paper was a small price for the bounties of power, while the former overlords could console themselves for the loss of the empire with the idea that they had bequeathed democracy to their former subjects. As with the entire myth of the British Commonwealth, it was a case of a delusion of omnipotence from the grave: though no longer on the spot to order the boys about, the former masters refused to believe that the instructions which they had left behind might be disregarded by the former pupils. The weakness of these farewell exhortations on democracy was that, until shortly before their departure, the colonial rulers had never tried to exemplify the practice of democracy, and were perfectly content with ruling in a strictly authoritarian manner.

The colonial administrations have created no foundations for government by consent. The few legislative assemblies functioned only within limited areas, had restricted powers and made no impact on the prevailing attitudes, which were much more influenced by the spectacle of the remote governor who brooked no opposition. However, even if the colonial governments had been making a serious attempt to implant democratic practices before bestowing democratic constitutions, it does not seem that they could have succeeded, for democracy is a tender plant which has thrived only in a few countries for longer than a brief period. In Britain representative institutions have existed for centuries, but the political rights were confined to a small minority until the Third Reform Bill in 1883, while universal suffrage dates only from 1918. Germany has had a democratic constitution only for fourteen years before the last war, and for eighteen years since, although the government of the Kaisers was extremely liberal, law-abiding and respectful of the rights of the citizen in comparison with what goes on today in ninety per cent of the members of the United Nations. Poland has had a democratic government for four years (1922–1926), and Russia at most during the six months between the fall of the Tsar and the Bolshevik seizure of power. France was lucky to have enjoyed democracy for over eighty years, but Spain has only had five. Most parts of the world, including such parts of Europe as Portugal and the Balkans, have never experienced it. So the Africans need not feel that the absence of democracy in their countries marks them as an inferior race. Democracy, to repeat, is a tender plant which requires very specific conditions.

The first necessary though not sufficient condition of democracy is the absence of mass poverty: although representative institutions with restricted franchise can function even if the majority live in painful misery. The conservatives, like Macaulay and Burke, who opposed the idea of extending the franchise, on the ground that such a step would lead to strife, chaos and eventually tyranny, were perfectly right on the assumption that the masses to be enfranchised would remain as poor as they were when these authors were writing. In all likelihood the democratization of the British constitution would have had such consequences had it not been followed by a general improvement in the standards of living, due to an acceleration of the growth of productive capacity, accompanied by a decline of the birth rate, while the ample outlets for emigration were also alleviating the pressure of the population upon the resources. The United States, where (apart from the South) democracy began with independence, have always had the highest standard of living in the world. The French were the first nation in the modern world to adopt the practice of restricting their progeny, and in consequence were able to attain the highest standard of living in Europe despite the relatively slow growth of their industry; and although the industrial workers had to suffer misery, the peasants were relatively prosperous and satisfied ever since the Revolution when they became the owners of their plots. Remarks along these lines could also be made about Switzerland and Scandinavia, and all the evidence shows that no political system which could be described as democratic has functioned for long in the midst of widespread misery. The explanation of this incompatibility is perfectly simple: democracy requires a consensus on the rules for investiture and exercise of authority, and the expectation that these rules will be observed. These conditions are impossible to maintain when goods are so scarce that people will resort to any means to obtain them unless restrained by force.

The basic political problem of the underdeveloped countries is usually approached from an unrealistic viewpoint: people argue about whether democracy can ensure economic progress or whether a dictatorship is needed for this purpose, whereas the real question is not so much whether dictatorship is desirable as whether it can be avoided. Such are the allurements of wishful thinking that even serious writers often do not realize that the proposition 'misery necessitates dictatorship' in no way entails the proposition 'dictatorship will cure the misery'. If the choice really lay between democracy and prosperity few reasonable people would advocate the former for countries where millions are starving, but in fact there is a natural affinity rather than incompatibility between democracy and prosperity: and the normal condition of mankind is to have neither.

Even an oligarchic consensus, however, does not exist in African states, owing to the ethnic heterogeneity of the new ruling élites. There are, it is true, historical examples of cohesive élites composed of individuals of variegated origins—such as the Catholic hierarchy or the ruling personnel of the Ottoman Empire—but these people were first uprooted, secondly indoctrinated and thirdly inserted into a rigid authoritarian framework. Only the military élites in Africa offer some analogies to this kind of situation, which may be a reason why in so many cases the military rule was the only alternative to disintegration of the post-colonial state. The loose structure of the civilian élites generates no centripetal forces which could counterbalance the centrifugal tendencies stemming from ethnic affiliations. The strong ties of ethnic and kinship solidarity, moreover, transform the fights between the factions of the ruling personnel into struggles between clans, tribes and ethnic groupings.

The fissiparous forces arising out of ethnic heterogeneity are not the only obstacle to an implantation of a representative form of government in Africa. Equally inimical is the set of circumstances which makes political power into the sole fount of wealth because the acerbity of the struggle for power depends largely on the number of people who will lose their jobs if the supreme authority changes hands. Another aggravating factor is the lack of alternative opportunities of making a fortune or even a bare living.

As the big business is in foreign hands, while the African capitalists employ only their kinsmen in lucrative positions, loss of office in an elected body or public service condemns the victim to destitution unless he succeeds in safeguarding a substantial part of the loot made while in power. Understandably, the knowledge that these are the only alternatives to continuing in office stimulates venality. There is a vicious circle here, because the habit of extorting bribes depresses the earnings of the smaller and more defenceless businessmen while augmenting the profits from access to power, which means that, as an avenue to prosperity, business becomes less and less promising in comparison with politics. The consequent deflection of ambitions into the channel of politics aggravates strife and reinforces the predatory propensities of the rulers and their henchmen.

The turn which African politics is taking—and in particular the role of the armed forces—does not appear so surprising if we view it in the light of what I propose to call the principle of naturalness of despotism. In order to obtain a better insight into the nature of despotism we should invert the problem: instead of attempting to explain the occurrence of despotism we should try in the first place to explain its non-occurrence. This inversion of the problem can be justified on the ground that despotism is the most natural form of government of large social aggregates— natural in the sense of the most probable, that is to say, requiring fewest specific conditions for its emergence and continued existence.

When we look at any example of a non-despotic system of government of a large political unit, we find that it contains an intricate mechanism of balance of power. Montesquieu's principle that despotism can be prevented only by a division of authority remains one of the greatest discoveries of sociology, and has fully withstood the test of time. As is well known, Montesquieu thought that it was the division between executive, legislative and judicial authority that made civil liberties possible in England. The writers on constitutional history usually say that he misunderstood the English system of government, but they are wrong because, although it is perfectly true that the division of authority has never been absolute in England, it was very real in comparison with the situation among Continental monarchies, let alone the despotisms of the East. Like all great discoveries, Montesquieu's idea was very simple: it amounted to the application of the old Roman maxim 'divide and rule' to the rulers themselves, thus converting it into 'divide your rulers in order not to be trod upon'.[1] True, he formulated it in terms which were too legalistic, but as reformulated by Mosca, who speaks of the balance of social forces, the principle is unassailable. However, the lack of a predominant centre of power does not need to produce a viable equilibrium: disintegration through strife or paralysis of the body politic are more probable outcomes. A political system based on an equilibrium of forces must generate conflicts, and at the same time contain them within narrow limits compatible with effective collective action. The naturalness of despotism is demonstrated by the fact that all non-despotic regimes were products of slow evolution under circumstances which can without much exaggeration be described as hot-house, and that every severe disturbance of their intricate structures caused a lapse into despotism. The movement away from despotism is long and laborious, whereas the movement towards it is quick and easy.

The recent military coups—in consequence of which the majority of the African states came under military rule—can be explained by the following factors:

(1). The complete lack of consensus on the rights to command and the duties to obey; due to (a) the very newness of the states, (b) their arbitrary frontiers and ethnic heterogeity, (c) the strangeness of the constitutions which have been foisted upon them by the departing foreign rulers.

(2). The bitterness of the struggle for the spoils, due to general poverty in combination with unrealistically high expectations and

---

[1] Although an intuitive awareness of this principle underlay the Greek and Roman customs of instituting plural executives (including war leaders), none of their political philosophers has formulated it explicitly. Polybius came quite near to it—with his idea of a mixed government based on a tripartite balance of power between the populace, the magnates and the ruler—but he was concerned with what was good for the power of the state in relation to its neighbours rather than with the rights and freedom of the citizens.

the fact that political office is the chief and often the only road to wealth.

(3). The weakness of the civilian supra-ethnic organization.

The first two factors rule out the democratic game and ensure that political authority be based on coercion, but do not predetermine whether the authoritarian rule will be civilian or military, which issue will depend on the relative strength of the two power machines. Where (as in Guinea, Tunisia and Ivory Coast, and perhaps Malawi and Tanzania) the ruler has been able to organize an efficient ruling party and secret police, he has been able to keep the soldiers at bay. Where this has proved impossible and the civilian power has been weakened by squabbles among the politicians (as in Nigeria) or the pursuit of ruinously unrealistic policies (as in Ghana), the men with the guns took over. Tanganyika and Kenya have been saved from this fate by the intervention of British troops, but it is by no means certain that they will not follow suit before very long together with other states in East Africa. As the examples of Ethiopia and Morocco show, however, a strong absolute or semi-absolute monarchy, with deep roots in tradition, can constitute an alternative to an efficient mono-party as a bulwark against military take-overs.

Though natural in the sense indicated above, military dictatorship need not constitute any solution, if by a solution we mean a step towards prosperity and peace, let alone democracy. As innumerable examples show, ranging from Ancient Rome to contemporary Latin America, pretorianism is one of the least stable forms of government, and not in the least immune to such diseases as corruption, inefficiency and internal strife. In the Congo there was hardly anything for the soldiers to destroy, but in Nigeria the same centrifugal forces which have ruined the civilian government are at the moment of writing rending asunder the military machine and bringing the country to the verge of chaos.

Latin American militocracy is radically different from the militarism which afflicted the European states and Japan. The Prussian militarism—to take the most developed representative of the species—entailed militocracy, that is to say, social and political predominance of the soldiers; but it exhibited few pretorianist proclivities, and its ideology was that of service for the cause of national grandeur. Although the Prussian officers insisted on occupying the places of honour in the society, and expected to live in a manner befitting their exalted position, they were not money-grabbers. Even when they became ideologically disoriented after the fall of the monarchy, and began to gravitate towards pretorianist incursions into politics, their chief aim was to prepare a renaissance of German military power. All militarists shared the assumption that what was good for the officer corps was good for the nation, but on the whole European militarism—and the same is true of Japan—was extroverted: it was oriented primarily towards fighting the foreigners, and only secondarily towards the task of protecting the social order from internal subversion. Naturally the relative emphasis varied according to the country and the time, but only in Spain was the protection of the privileged classes against internal dangers the chief function of the army.

In Latin America there was only one example of militarism in the sense of regimentation of the entire population for the purpose of waging war: it was Paraguay in the middle of the last century, which its dictators Francia and the two Lopez made into one great military camp. Its militarization was so thorough that this sparsely populated country could wage war for several years against the combined armies of Argentina, Brazil and Uruguay; and by the time it was defeated it had lost three-quarters of its male population. However, the Paraguayan war was the biggest in Latin American history. The war of the Pacific (Chile against Peru and Bolivia) was decided by naval engagements and the land forces were small.

The forces which defeated the royal armies during the Latin American Wars of Independence were motley crowds of volunteers, few of whom shared Bolivar's and San Martin's concern for liberal ideals. Hatred for the Spanish colonial administration was the only common conviction, and once the Spaniards departed, there was no ideology left—not even nationalism, because the embryonic nations were too amorphous to inspire deep devotion. So there were sparsely populated territories over which roamed war bands that here and there succeeded in establishing a semipermanent sway. These recruited *ad hoc* bands had neither standard equipment nor proper uniforms, nor systematic training. Their commanders—no drawing-room officers but rough men who had proven their mettle on battlefields—fought for division of the spoils and personal supremacy unhampered by ideological considerations.

When there is neither political consensus nor institutions permitting orderly governance, the only possible way of governing a country is by force. Thus pretorianism was the inevitable consequence of the wars of liberation which destroyed the colonial administration together with its ideological basis, without putting in its place any institutions which could command general assent. And as arbiters of politics the military arrogated to themselves extravagant privileges and consumed a ruinously large share of national wealth.

The military function has become introverted in the Latin American republics; with few opportunities to fight for their countries, the soldiers remained preoccupied with internal politics and the search for personal and collective advantage. One of their most striking characteristics is their unideological and mercenary outlook.

During the first two or three decades after Independence the armies functioned as agencies of social reshuffling rather than as props of the established order: the rich were not especially attracted to the life of danger and exertions, and as in the hour of danger military prowess mattered more than family connections, many men of humble origins became commanders of armies and eventually presidents.

When life became more regular and the republics acquired some shape, the armed forces began to acquire the character of regular armies, particularly in the better-ordered countries. This process was slow, and it was not until towards the end of the nineteenth century that the armies of the more progressive republics came to resemble their European counterparts. Chile led the movement of professionalization helped by a German military mission, and by 1880 Chile, Argentina, Mexico and Brazil had regular national armies, although in the more backward countries like Guatemala and Bolivia the armies still retained the character of marauding bands. The movement towards professionalization was stimulated by the coming of the new weapons and by conscious imitation of the European models. Artillery in particular required more elaborate organization of supply and greater technical preparation, and made it more difficult to seize power by organizing a band of *gauchos* in the pampas and then marching on the capital. The revolutions did not cease, but thenceforth they usually were fights between different factions of the officer corps, and for this reason they were less drawn out and bloody.

As, with the abatement of warfare in the more progressive countries, prowess in combat began to matter less and military life became less arduous, the privileged classes reserved for themselves the higher military posts. In consequence the officer corps became in most countries an appendage of landed aristocracy. The rank and file were naturally recruited among the poor. Even in those countries which had laws prescribing universal service, in practice only the peasants were liable to be conscripted. Usually illiterate rustics were preferred, as they were regarded as more dependable than the city dwellers, who might have their own views about politics.

Integration of the officer corps with the landowning aristocracy, in combination with consolidation of civilian political institutions, produced a certain decline of pretorianism in the second half of the nineteenth century. Apart from Chile, where this process was already accomplished by 1830, government became predominantly civilian in Argentina by 1860, in Uruguay by 1890, in Colombia by 1900. In Brazil the continuity of the monarchy prevented the worst excesses on the Hispano-American pattern and, although the military frequently caused a considerable amount of trouble, it was only during the brief period after the abolition of the monarchy, and before the aristocracy had organized republican institutions, that pretorianism dominated Brazilian politics.

The trend away from pretorianism was limited to the economically more progressive countries: Central America, Ecuador, Bolivia and Paraguay were not affected. In Mexico Porfirio Diaz crushed his rivals and succeeded in maintaining order for more than three decades, but this was a military dictatorship.

The influence of the soldiers upon Latin American politics receded until the twenties of the present century. The region's prosperity was growing, and the changes in the social structure did not yet call into existence mass movements offering an open challenge to the established order, except in Chile, ruined by the collapse of the market for nitrates, and of course Mexico.

As Edwin Lieuwen says in *Arms and Politics in Latin America*, pp. 122 ff, 'at the time of World War I, the fraction of the total area and population that was dominated by the military was declining, and by 1928 only six Latin American countries, containing but 15% of the total population, were ruled by military regimes. Then, abruptly, following the onset of the world depression in 1930, there occurred a striking relapse into militarism' —i.e. pretorianism.

The armed forces re-assumed the political prominence which they had in the earlier parts of the nineteenth century, but their role was now much more complex and ambiguous, owing to the entry of new forces into the political arena, and to the changes in the social roots of the officer corps. As the urban middle classes grew in numbers and importance, they began to send some of their sons into the army, and these men did not feel the same attachment to the cause of the Church and the landed aristocracy.

The struggle for the spoils was, of course, nothing new. What was new in relation to the preceding decades was the willingness of the younger officers to enlist the support of the lower classes, frequently resorting to crass demagogy. In this respect, some of the more radically disposed officers, like Peron in Argentina or Arbenz in Guatemala or Busch in Bolivia, resembled Bolivar in the earlier stages of his career and other populist generals of the wars of independence, like the Chilean Bernardo O'Higgins. But in contrast to these early populist *caudillos*, the recent populist pretorians rose to power with the backing of officers' secret societies such as the Lodge of the Holy Cross in Bolivia, the Group of United Officers in Argentina or the Patriotic Military Union in Venezuela.

The fruits of the reforms sponsored by the populist officers have not been very impressive. Many of them resorted to demagogy as a tactical weapon in the struggle against the established rulers, but once in power they betrayed their lowly supporters, and devoted themselves mainly to building private fortunes. Such was Ibañez in Chile—the first demagogic military dictator in the twentieth century in Latin America outside Mexico—such also was Sergeant Batista in Cuba, who, at the beginning of his political career, had the reputation of a communist revolutionary. Other pretorianist reformers have been overthrown by the officers (often older and senior in rank) devoted to the defence of the upper classes. This was the fate of Colonel Rafael Franco of Paraguay, of

POSTSCRIPT TO THE SECOND EDITION

Colonel Arbenz of Guatemala, of General Rojas Pinilla in Colombia (although his populism was purely verbal), of Major Osorio in San Salvador, and of Colonel Busch and Major Villaroel in Bolivia, who were both murdered. Peron could also be included in this category, but his case is more complicated.

The armies of Latin America cannot fight each other, because in virtue of a multilateral treaty the U.S.A. would come to the defence of the invaded country. The security of the hemisphere against an invasion from outside is also guaranteed by the U.S., and should their forces prove inadequate for this purpose, the balance could not be redressed by the armies of Latin American republics, which lack modern equipment and training; particularly as officers accustomed to insubordination and intrigue would be of little value in war. The events in Colombia have shown that such an army cannot even cope with guerrillas.

When viewed as props of the social order the Latin American armies are highly defective and unreliable instruments. The conscripts cannot be relied upon not to go over to the rebels, and even the loyalty to the present regimes of some of the officers is doubtful. For the purpose of suppressing internal subversion a smaller but well-trained police force, like the Chilean *carabineros*, is much more reliable than a conscript army; and it seems that conscription is retained in so many republics chiefly in order to swell the numbers so as to provide jobs for officers. The armed forces behave as cancerous growths which, instead of performing any service to the social organism, only harm it.

According to U.N. sources, the military establishments in Latin America in 1956 were as follows:

	Army	Navy	Air Force
ARGENTINA	107,000	21,500	19,000
BOLIVIA	15,000	600	2,000
BRAZIL	90,000	8,000	9,200
CHILE	20,500	8,000	13,000
COLOMBIA	10,000	1,500	200
COSTA RICA	has only a small police force		
CUBA	19,000	2,000	2,400
ECUADOR	no data		
EL SALVADOR	6,000	400	500
GUATEMALA	21,000	1,000	400
HAITI	4,500	nil	400

	Army	Navy	Air Force
HONDURAS	2,500	nil	1,200
MEXICO	41,000	2,500	3,500
NICARAGUA	10,000	nil	1,300
PANAMA	only a police force		
PARAGUAY	5,800	400	nil
PERU	10,000	2,500	5,000
DOMINICAN REPUBLIC	3,500	3,000	2,000
URUGUAY	3,000	1,450	200
VENEZUELA	10,000	2,240	5,000

This list shows that the armed forces of Latin America are too large for police duties, but too small for waging war. Apart from Cuba since 1961, only Argentina and Chile could resist a minor invasion.

In relation to their size and poverty of equipment these forces are exceedingly expensive, owing to disproportionately large numbers of high-ranking officers, both on active service and retired, and their high emoluments. Argentina, for instance, has as many generals as the United States.

The African armies are extremely small in comparison, despite their prominent political role. Between the Sahara and the Zambezi there are only 130,000 soldiers, almost entirely infantry or cavalry without heavy weapons. None of the tropical African states has an air force or a navy capable of effective action. The armies of the North African states, on the other hand, add up to over 700,000 men and possess tanks, jet combat planes, rocket detachments and anti-aircraft units. The South African army numbers over 100,000, and Rhodesia has armed 40,000 men. The Portuguese army in Africa amounts to about 100,000. This means that the white southern armies have an almost 2: 1 superiority, apart from being far better organized and armed.

None the less, in 1965, the states of tropical Africa spent on arms 2,300 million dollars, which amounts to about twenty per cent of their estimated joint national incomes and double all industrial and agricultural investments.

In his book *El Militarismo* (Mexico 1959), Victor Alba calculates that military expenditure per head in Argentina is four times greater than in Mexico; in Chile almost five times greater; in Cuba (pre-Castro) six times; and in Venezuela twelve times. This accounts for the unrivalled rate of economic growth in Mexico.

In 1958, according to this source, the situation in this respect was as follows:

	Percentage of military expenditure in the budget	Military expenditure per head in dollars
ARGENTINA	25·5	8·82
BRAZIL	27·6	4·13
CHILE	22·3	10·80
COLOMBIA	16·9	2·30
CUBA	16·9	12·31
GUATEMALA	8·2	2·63
MEXICO	10·4	2·12
VENEZUELA	11·2	25·91

Militocratic appropriation of wealth constitutes the most important form of parasitism in Latin America. The wastage of wealth occasioned by it is even greater than that caused by absentee landlordism, because at least a part of the profits of the landlords is put into socially useful investments.

Though in varying measure, all the armies of Latin America are debilitated by corruption and indiscipline. If we tried to rank them in this respect, we would probably have to put the Chilean army at the top as the least addicted to these vices, and the Argentinian at the bottom—at least at the moment. Indeed, as far as Argentina is concerned, it seems that the venality and indiscipline of its soldiers constitute the most important source of its troubles. In most African countries the situation in this respect is far worse, and complete breakdowns of order have been common occurrences.

Fidel Castro's revolution in Cuba has demonstrated how weak an army can be, even if well armed and in possession of the requisite technical skill, if it is corroded by graft. His partisans were able to defeat the much more numerous and better-armed professional army of Batista, because the soldiers were engaged in peculation and fraud, and were completely devoid of any sense of honour, duty or even solidarity. The examples of modern totalitarianisms show that a well-organized machine of terror cannot be be overthrown from within so long as it remains austere, ideologically committed and led by capable men, but none of these attributes pertains to the Latin American armies of today. It follows that as a bulwark against communist subversion they are not very dependable, although this does not mean that military dictatorships have no chance of establishing and maintaining themselves for a considerable time; but, being incapable of solving their countries' basic problems, and even aggravating them, they can merely stave off the danger.

6. NUCLEAR WEAPONS AND THE NEW MARTIAL VIRTUES[1]

It is hardly worth while to discuss what might happen to mankind in consequence of an all-out nuclear war consisting of indiscriminate mutual destruction. The most likely outcome of such an eventuality—the extinction of humanity—does not lend itself to further sociological speculation. The possibility that the social evolution of mankind might begin anew from the level of primitive existence, if there are some survivors, is more interesting, although the chances seem to be that, owing to the exhaustion of the ores which could be mined with the aid of primitive techniques, mankind would never again be able to transcend the level of the Stone Age, apart from the well-known possibility that radioactivity might cause biological degeneration of the species.

The shrinking of the world and the fantastic and still increasing deadliness of the weapons rule out the possibility that the present arms race may go on for a very long time, which leaves two possibilities: conquest, or general and genuine agreement.

The prospects of a genuine agreement are not encouraging. Human nature is such that bickering comes naturally, and nothing is more common than examples of the propensity to 'cut off one's nose to spite one's face'. This general human frailty is dangerous enough, and the discrepancy between mechanical ingenuity and moral backwardness may very well prove fatal to mankind. This situation is connected with the fact that whereas the technical level is fixed by the achievements of the most gifted inventors, the ethical level is determined less by the most benevolent of men than by their opposites, because of the way in which the processes of selection for positions of authority favour ruthless power-seekers. This seems to have always been so, but it is worth

[1] Reprinted and abridged from *Elements of Comparative Sociology* (Weidenfeld & Nicolson, London 1964); published in the U.S.A. by California University Press in 1965 under the title *Uses of Comparative Sociology*.

noting that the diminution of role played by inheritance has brought no improvement in this respect. When thrones and dignities were hereditary they often fell to imbeciles or blood-thirsty madmen, but on occasion they came into the hands of benevolent men or women. When the posts of command are thrown open to competition, idlers and imbeciles have no chance, but neither have the gentle nor those with too many scruples, whilst bloodthirsty madmen are by no means excluded. A socio-logical generalization can be proposed that the less determined is the selection by factors other than the sheer ability to manipulate men, the more ruthless and astute will be the rulers. This socio-logical theorem explains why American presidents and British prime ministers could be outmanœuvred by Soviet rulers. It also explains the relative mildness of European kings as compared with the monarchs of the East, bred in polygamous households, amongst numerous half-brothers whom they had to kill in order to mount the throne. To win a competitive prize one must desire it very strongly, and therefore nobody is likely to have power who does not crave it. For this reason, the issues on which the survival of humanity depends will in all likelihood continue to be treated as gambits in the game of power-seeking.

Imposition of hegemony puts fewer demands on human nature, as it could result from a free play of very common propensities. Whether an attempt in this direction would in fact lead to the establishment of a world government or to a holocaust would depend on the situation in military technology and on chance. I shall say nothing about the latter element, and only a few words about the former—solely in order to dispel the common pre-conception that a victory in a future war is out of the question. It is true that as things stand now an all-out war can benefit nobody, but we must not assume that this state of affairs is immutable. Some startling inventions might make indiscriminate blasting with hydrogen bombs futile or even impossible. Some combination of radar and television might enable people to see any point on the globe or even inside it. Some rays might be invented which kill or paralyse living creatures within the area on which they are directed, or even impede the functioning of mechanisms by creating magnetic fields, using anti-matter or something else that cannot yet be imagined. The argument that such things can never be done carries little conviction, because it has been found to be wrong so many times before. It is clear, however, that a victory in a future war can only be the fruit of superior technical inventiveness, and could never result from the mere multiplication of the armaments existing at present.

Before proceeding to examine the social implications of a race in military technology, I should like to make an aside and point out that there are few reasons for believing that a world government—an ideal earnestly desired by generations of humanitarian political thinkers—would in fact be so ideal, even if it came into existence through negotiation instead of conquest. Experience with existing international organizations demonstrates the law of the least common denominator, which Gustave le Bon formulated in order to account for the frequently observed fact that people behave worse in crowds than individually. The level of probity, efficiency and enlightenment in an international organization is about the same as in the administrative machines of its member states with the lowest scores on these points. This is in no way surprising, because probity and efficiency can be attained only if at least the persons in key positions act in accordance with commonly accepted principles, and how can this occur in an organization whose personnel have no common convictions—whose declared purposes are merely declared but not seriously felt. It might be said that these features are the consequences of the extremely federalist—nay, amorphous and acephalous—character of the present international organizations, and that they would disappear in a real world government. Even if we concede this point —which in fact carries little conviction outside the possibility of universal hegemony—two indelible vices of a world government would remain. The first would be its size, because largeness (as has been known since Plato and Aristotle) renders any form of control from below less effective and favours abuse of power. The second indelible vice of any world government would stem from the mere fact that it would be the only sovereign government— a government from which there would be no escape. To appreciate the portent of this fact we must remember the enormous importance which possibilities of escaping had not only for alleviating suffering but also for encouraging resistance to tyrannies. We must also bear in mind the prominent role of exiles in spreading the spirit of freedom and, in particular, the spirit of free thought. Living under a despotism has been the normal fate of mankind ever since large states arose, and intellectual and moral progress has been possible only because from time to time little evanescent islands of freedom emerged which gave to a few generations of their favoured sons the opportunity to think and speak freely. Even if it avoided excesses of tyranny, an effective world government would most likely suffocate the germs of intellectual and moral progress by the dead weight of bureaucratic conformity. This prospect might be better than that of a a holocaust, but is hardly very entrancing. It must be remembered, moreover, that in the past governments showed concern for their subjects mainly when they needed their co-operation in fighting foreigners. In a world state there would be no threats from outside to countervail the normal tendency of rulers to segregate themselves from the ruled, and to ill-treat them.

In order to be militarily decisive an invention (or rather a series of inventions) must be kept secret. In this respect the advantages of totalitarian regimes are obvious. In the U.S.A. even the army is not very good at safeguarding secret information. The craving for publicity is so deeply embedded in the American tradition, that it could not be extirpated without radical changes in the constitution and the mentality of the people. The executive might have to be freed from supervision by Congress, so as to prevent investigating committees from compelling generals and officials to make public the information which spies could obtain only with a good deal of work. Freedom not only to travel abroad but even to move without restrictions within the boundaries of the state facilitates spying enormously, and makes it quite impossible to keep secret the location of sizeable installations. The security of tenure of official posts, the adherence to the principle of treating a person as innocent until his guilt is proved, and the rule of law in general, all constitute serious handicaps in counter-espionage. It does not follow, of course, that these handicaps cannot be sources of strength in other ways, but the fact remains that they give a very substantial advantage to the totalitarian state in the race in military technology.

Liberal states labour under another serious disadvantage in the contemporary technological race. In any of them a scientist who undertakes to work on a military project has to forgo freedoms which all other civilians enjoy: unlike his colleagues who are occupied with pure science and teaching, he is under continuous observation, his private past is open to inspection, he has to be careful about the friends he makes and so on. In contrast, his counterpart in Russia or China loses nothing by taking up military work, because he would be subject to all these restrictions in any walk of life. He may even gain some freedom because, being a valuable worker, he is less likely to be thrown into prison for a fictitious or insignificant shortcoming. This means that a totalitarian state has much less difficulty in attracting the best brains to military technology. Another advantage stems not so much from the totalitarian character of the Russian and Chinese regimes as from the poverty of their subjects: the difficulty of attaining a fairly comfortable existence ensures that everybody is extremely susceptible to material inducements, and thus facilitates the direction of brain power, which in any case is easier in an authoritarian than in a libertarian state.

The requirements of secrecy and direction of brain power foster certain features normally included in the concept of totalitarianism. This does not mean, however, that all the features of totalitarianism as we know it enhance the chances of obtaining technico-military preponderance. The old-fashioned totalitalianism of Hitler and Stalin was perfectly adapted to the needs of mass warfare. Cannon-fodder, as well as man-power needed for production, had to be harangued, doped by rituals, and terrorized at the same time. Now that the military usefulness of the masses diminishes, the need to control them so stringently decreases too. It would seem, therefore, that obligatory enthusiasm, mammoth parades and other related features of the rules of Stalin, Mussolini and Hitler have outlived their military usefulness, though the possibility still remains that they might survive for other reasons.

On the whole it seems that nuclear weapons favour a kind of moderate and rationalized totalitarianism; without extravagant regimentation of the masses and ritual follies, without absurd incursions by the rulers into the realms of natural sciences, without crazy doctrinarianism, but with thorough and systematic control over the movements and work of the inhabitants, and absence of freedom to form associations, let alone to criticize the regime.

There is another very general aspect of the impact of nuclear weapons on social life: I do not mean the often noted phenomenon of the widespread feeling of futility of existence, induced in many imaginative persons by the imminence of the holocaust, but the

e introduction of these weapons constituted the last
ne devaluation of martial virtues which began with the
tion of gunpowder. When firearms first appeared they were
nerally condemned on the ground that they permitted a
dastardly weakling to kill a man of strength and valour. Now we
reach the stage when a sickly woman scientist suffering from an
anxiety neurosis may constitute a far greater military asset than
thousands of tough and fearless soldiers.

The chief military virtue is nowadays neither physical endurance nor pugnacity, but technical inventiveness; and in view of the
fact that military virtues are usually extolled, we should expect the
spread of the cult of this quality. The ideal which is most useful
from the military point of view is that of a man or woman who
invents what is demanded without asking why, and in fact this
ideal is energetically inculcated throughout the breadth of the
Soviet Empire. On the other hand, in the West (above all in the
U.S.A.) we find the opposite: cheap literature as well as mass
media propagate the cult of brainless toughness combined with
disdain for 'egg-heads'. There are two possible explanations
(which do not exclude each other) of this trend which may well
prove fatal to the West. One is that it is simply engineered by the
advertisers, who dislike people who think too much and do not
buy without asking why, and for this reason foist upon the public
the ideal of a brainless tough and a dumb doll. The other explanation of the present cult of male toughness is that it is a spurious
compensation for lost reality. When men carried swords and used
them habitually, and when life was altogether dangerous, they
found nothing repugnant in powdering themselves, wearing frills
and observing the punctilia of courtesy. In contrast the young
men of today, pampered at home, protected by the police, prepared by their teachers to be smooth organization men, live in
dread of being taken for 'chickens' and find in rudeness and
'tough speech' the only opportunities for proving their manhood.
There is a curious historical parallel of this phenomenon. In
ancient Sparta the training in endurance, designed to produce
future warriors, became extravagantly cruel when it lost its usefulness—when Sparta came to be just one of many little towns of the
Roman Empire. Just as the contemporary cult of toughness, this
example shows how a feature of character may come to be stressed
most when it entirely loses its original function.

POSTSCRIPT TO THE SECOND EDITION

The use of mass media of communication in such a way as to foster the cult of violence, selfishness and stupidity, instead of teaching civic virtues, may very well become the chief cause of the defeat of the West, which may take place even without an outright war; because whatever the evils of doctrinarian propaganda, it must be admitted that in the communist states mass media are not used for the purpose of extirpating all the virtues which are necessary for the strength of any state.

Nuclear weapons might bring mankind to its doom or they might turn out to be a boon, but they certainly prevent mankind from continuing on its normal course of customary brutality: men will have to treat each other better or they will all perish. By making war suicidal even for the rulers, atomic weapons took away from them the opportunities of pushing others into carnage for the sake of their own amusement and glorification. Despotism still represents a great danger to peace because of the possibility that a madman might reach the position in which he could not be restrained from bringing everybody to destruction. A rational despot, however, cannot nowadays be excessively bellicose.

Countless covenants of peace and condemnations of war have proved to be of no avail against the evil propensities of rulers, but there is a slight chance that the immediate danger to their lives might be more effective. It all hinges, however, on whether they will behave rationally.[1]

Apart from general impediments to rational behaviour stemming from the apparently ineradicable waywardness of human nature, irrationality has been assiduously cultivated by various institutions, above all, the armies. Inculcation of blind obedience and of readiness to die without asking why, and elevation of these habits to the status of the highest virtues, cannot fail to propagate irrationality. With the devaluation of the traditional martial virtues, rationality may come to be more appreciated. To be exact, this process has been going on for a very long time—in fact at least since the invention of firearms—but until now its pace has

---

[1] However, the wars in Korea and Vietnam constitute a new type of warfare limited by the tacit or even explicitly agreed conventions along the lines of 'I shall kill your pawns and you try to kill mine, but it would be too dangerous for us to start hitting one another'. This kind of warfare can assuage the bellicose propensities of rulers and mobs, and liquidate the demographic surpluses, without endangering either the rulers themselves or the cores of their domains. Other variants of this type of warfare will no doubt emerge in due course.

been exceedingly slow and there have been numerous relapses. In contrast, the change in this respect wrought by the advent of atomic weapons promises to be quick and radical. It will probably diminish the chances of irrational types attaining positions of power, and thus reduce the risks of war. Bureaucratization also acts in the same direction, as it favours the ascent of calculating manipulators rather than of fiery demagogues or strong-arm men. It was not goodwill towards the capitalists but cold calculation of consequences which has persuaded the Soviet rulers of the advantages of coexistence. There is some assurance of peace in the fact that in the struggle for promotion within the Soviet system calculating manipulators have an advantage over fanatics.

7. A NOTE ON NEOLOGISMS

Among the neologisms introduced in the first edition, 'bookay' and 'bookayan society' deserve complete oblivion. Although it sounds better, 'polemity' is also unnecessary, as it can be replaced without any loss of meaning by 'militancy'—one of Herbert Spencer's favourite words. The expression 'pheric distance' does not seem to be particularly useful either. The term 'biataxy'— defined as the degree to which the structure of a society is determined by the use of violence—can be replaced by 'co-ercivity'.

The names of the six ideal types of military organization can be made self-explanatory if we make the following substitutions:

homoic can be replaced by	polis-type
masaic – – –	– tribal
mortasic – – –	– restricted professional
neferic – – –	– widely conscriptive
ritterian – – –	– feudal
tallenic – – –	– inarticulate sub-tribal

Pareto used 'ophelimity' instead of 'marginal utility'. It is doubtful whether it is useful to employ it in the sense of generally desired good things of life such as wealth, power and glory—as is done in the first edition of the present book. Perhaps the best generic term for wealth, power and glory is 'invidious values'.

Bibliography[*]

IT is impossible to enumerate all the books and articles which contributed towards the coming into being of the present work. I shall mention, therefore, only those works which are directly relevant. BERTRAND RUSSELL's stimulating book on *Power* (London 1936) set my thought on the path which led to the present book.

In the conceptual analysis of social phenomena I have been greatly influenced by my one-time teacher CZESLAW ZNAMIEROWSKI, whose writings served me as a model of clarity and precision; they constitute, in my opinion, the highest achievement of the formalistic approach to sociology; the most important of them is *Prolegomena do Nauki o Panstwe*, (Warsaw 1930).

My debt to MAX WEBER is enormous. Without the basis provided by his works I could never have written this book. He was very much alive to the sociological importance of military organization and touches on it even in his *Religions-soziologie* (3 vols., Tübingen 1922) (English translation in course of publication by T. Parsons). His views on this problem are stated most explicitly in the wonderful monumental treatise *Wirtschaft und Gesellschaft* (Tübingen 1922) (English translation of Part I, *Theory of Social and Economic Organization*), and in the masterly essay 'Zur ökonomischen Theorie der Antiken Staatenwelt' reprinted in *Gesammelte Aufsatze zur Sozial-und Wirtschaftsgeschichte* (Tübingen 1924).

I have drawn quite a number of ideas from HERBERT SPENCER's *Principles of Sociology* (3 vols, London 1876–96). His treatment of Political Institutions remains extremely valuable. GASTON BOUTHOUL's *Huit Mille Traités de Paix* (Paris 1948) and *Cent Millions de Morts* (Paris 1946) rescued the present enquiry from an impasse of apparent contradictions into which it got at a certain stage. They constitute an amazingly penetrating examination of the role of demographic factors in social conflicts.[1] Another excellent and indispensable study of similar problems is A. & E. KULISCHER, *Kriegs-und Wanderzüge* (Berlin 1932). I found very useful *A Study of War* by Q. WRIGHT (2 vols., Chicago 1942), a careful and painstaking examination of various theories about the causes of war, and J. WISSE, *De Strijdende Maatschappij* (The Hague 1948), a convenient short examination of various theories of conflict. R. S.

* Only works used in preparing the first edition are indicated. A number of books on civil–military relations have appeared since then, the best general treatment being in my opinion *The Man on Horseback* by S. E. Finer, Pall Mall Press, which also contains a bibliography of recent books. Among the books dealing with special areas the best seems to me Edwin Lieuwen's *Arms and Politics in Latin America*, quoted in the text of the Postscript to the present edition.

[1] Since this was written his treatise has appeared: *Les Guerres*. It is certainly the best work on causes of war.

BIBLIOGRAPHY

STEINMETZ, *Soziologie des Krieges* (Leipzig 1929), though too propagandist to deserve its title, contains some instructive views, particularly on biological selection and determinants of ferocity. A great deal of material concerning primitive warfare can be found in T. S. VAN DER BIJ, *Ontstaan en Eerste Ontwikkeling van den Oorlog* (The Hague 1929).

The second volume of FRANZ OPPENHEIMER's *System der Soziologie: Der Staat* (Frankfurt 1929), though it contains some untenable theories, is extremely important.

I have leant heavily in many respects on the works of PITIRIM SOROKIN, though I by no means accept the mystical superfluities contained in his later books. The works in question are: *Contemporary Sociological Theories* (New York 1928), *Social Mobility* (New York 1927), *Man and Society in Calamity* (New York 1942), *Social and Cultural Dynamics*, vol. 3 (4 vols., New York 1937–41), *Society, Culture and Personality* (New York 1947). Chaotically written but still unique in anthropological literature for the breadth of scope and the originality of views is RICHARD THURNWALD's *Die menschliche Gesellschaft in ihren ethnosoziologischen Grundlagen* (5 vols., Berlin 1930–5). It is particularly instructive on the phenomenon of conquest. Some of my factual data are also drawn from this work.

I am greatly indebted to GAETANO MOSCA, particularly for his discussion of the factors which determine the amount of military influence in politics, contained in *The Ruling Class* (New York 1939), which is, I think, the most illuminating treatise on politics ever written. I also used his *Histoire des Doctrines Politiques* (Paris 1936) and *Partiti e Sindicati* (Bari 1949), a reprint of articles. BERTRAND DE JOUVENEL's brilliant *Du Pouvoir* (Geneva 1947) helped me greatly in many ways, particularly in problems concerning the inner structure of hierarchies and their relations to one another. ALEXANDER RÜSTOW's penetrating analysis of the phenomenon of domination—*Ortsbestimmung der Genenart*, Zürich 1950 —enabled me to improve the manuscript on some points.

Traces of the influence of the following works can also be found on the preceding pages: V. PARETO, *Traité de Sociologie Generale* (Paris 1917) and *Les Systemes Socialistes* (2nd ed., Paris 1926); R. MICHELS, *Political Parties* (London 1916); G. SIMMEL, *Soziologie* (Leipzig 1908); J. M. ROBERTSON, *The Evolution of States* (London 1912); R. A. ORGAZ, *Ensayo sobre las Revoluciones* (Cordoba 1945); R. M. MACIVER, *The Web of Government* (New York 1947).

Among the works of the old masters those which inspired me most are: ARISTOTLE's *Politics*; IBN KHALDUN's *Prolegomenes Historiques* (trans. de Slane, Paris 1934–8, 3 vols.); MONTESQUIEU's *De l'Esprit des Lois*; MACCHIAVELLI's *Prince*; *The Book of Lord Shang* (trans. Duyvendak,

London 1928); KAUTILYA's *Arthasastra* (trans. Shamasastry, Mysore 1929); MALTHUS' *Essay on the Principle of Population*.

Quite a number of propositions expounded above have been arrived at by generalizing and modifying the interpretations of particular events given by some historians. Thus my theory concerning the influence of monetary circulation on social structure has been formulated in this way on the basis of Henri Pirenne's interpretation of the history of medieval Europe. I am similarly indebted to Rostovzeff, Gordon Childe and Fustel de Coulanges. JACQUE PIRENNE's brilliant interpretative history *Les Grands Courants de l'Histoire Universelle* (Geneva 1944–8, so far 3 vols.) is a mine of sociological ideas. R. TURNER's *The Great Cultural Traditions* (New York 1941, 2 vols.) is an excellent work of similar nature, narrower in scope but more reliable.

I have taken a few hints from A. J. TOYNBEE. Although the theoretical framework of his *Study of History* (6 vols., London 1935) is tautological and, as far as I could discover, completely nebulous, and the factual data mostly superficial and sometimes quite misleading, yet one can find there a number of very interesting suggestions and pieces of information.

The only book which is explicitly devoted to the question of the influence of military organization on the life of societies is MAX JÄHNS, *Heeresverfassungen und Völkerleben* (Berlin 1885), which, however, does not go deeply into sociological problems. The standard military histories which I used are ROLA-ARCISZEWSKI, *Sztuka Dowodzenia* (Warsaw 1936); J. ULRICH, *Kriegswesen im Wandel der Zeiten* (Potsdam 1940); HANS DELBRUECK, *Geschichte des Kriegkunst* (3 vols., Berlin 1900ff) (the following volumes written by others are of little value). The most sociologically minded general military history that I know of is P. SCHMITTHENNER, *Krieg und Kriegfuehrung im Wandel der Weltgeschichte*, (Potsdam 1930), which does, however, contain some naïve ideas. The information about Russia, the Mongols and the Islamic countries can be found in FERDINAND LOT, *L'Art Militaire et les Armées au Moyen-Age* (2 vols., Paris 1946). PAWLIKOWSKI-CHOLEWA, *Die Heere des Morgenlandes* (Berlin 1940) is the only book dealing with Asiatic countries generally. Otherwise, military histories of those countries do not exist in European languages. Scattered pieces of useful information can be found in B. LAUFER, *Chinese Clay Figure* (Chicago 1914); P. HORN, *Das Heer—und Kriegwesen der Grossmoghuls* (Leiden 1894); G. OPPERT, *Weapons, Army Organization and Political Maxims of the Ancient Hindus* (London 1880). A work of unequal value but containing much information not to be found anywhere else is LEO FROBENIUS, *Weltgeschichte des Krieges* (Hanover 1903). Problems of modern Europe are treated in A. VAGTS, *A History of Militarism* (New York 1937). Some books on general, legal, economic or social history give more information about sociologically relevant features of military organization than military histories.

My data concerning western Europe and the Graeco-Roman world can be checked in any general history series, like *Peuples et Civilisations*, or the *Cambridge Histories* (Ancient, Medieval, Modern) or *Propylaen Weltgeschichte*. I shall enumerate, therefore, only the books which I found particularly illuminating from the sociological point of view. These are: G. SEIGNOBOS, *Essai d'une Histoire Comparee des Peuples de l'Europe*, (Paris 1938); FUSTEL DE COULANGES, *La Cite Antique* (28th ed., Paris 1922); the role of military factors is underlined in J. HASEBROEK, *Griechische Gesellschaft-und-Wirtschaftsgeschichte* (1931); R. V. PÖHLMAN, *Geschichte der sozialen Frage and des Sozialismus in der antiken Welt* (3rd ed., Munich 1925); G. GLOTZ, *La Cité Grecque* (Paris 1928) (English translation, *Greek City*, 1931); GORDON CHILDE, *What Happened in History* (Harmondsworth 1942); W. W. TARN, *Hellenistic Civilization* (2nd ed., London 1930); M. ROSTOVZEFF, *The Social and Economic History of the Roman Empire* (Oxford 1926) and *Soc. and Ec. Hist. of the Hellenistic World* (Oxford 1941); L. HOMO, *Roman Political Institutions* (London 1929); F. LOT, *La fin du Monde Antique* (Paris 1927) (English translation, *The End of the Ancient World*, 1929); MARC BLOCH, *La Societe Feodale* (Paris 1939); H. MITTEIS, *Der Staat des Hohen Mittelalters* (3rd ed., Weimar 1948); HENRI PIRENNE, *A History of Europe* (London 1933); WERNER NÄF, *Die Epochen der neuren Geschichte* (2 vols., Aarau 1945).

On Byzantine History: CHARLES DIEHL, *Byzance* (Paris 1919) and *Les Grands Problemes de l'Histoire Byzantine* (Paris 1943); G. OSTROGORSKY, *Geschichte des Byzantinischen Staates* (Munich 1940). The most comprehensive picture of Byzantine society and civilization can be found in L. BREHIER, *Le Monde Byzantin* (3 vols., Paris 1947–51). Very interesting are W. E. D. ALLEN, *A History of Georgian People* (London 1932); and H. PASDERMADJIAN, *Histoire de l'Armenie* (Paris 1949).

For data on the history of Poland and other Slavonic countries I relied on works available in Slavonic languages only.

Among short histories of Spain I found most useful: A. PALOMEQUE, *Historia de la Civilizacion e Instituciones Hispanicas* (Barcelona 1946), although RAFAEL ALTAMIRA's *Historia de España* (4th ed., 4 vols., Barcelona 1928) remains unrivalled.

The list of books from which I drew my information about Asiatic countries must be much fuller because most books on oriental history are just inventories of incidents, and those that give any information about societies are very few.

On the Ancient Near East I found most useful: C. DAWSON, *The Age of Gods* (London 1928); A. MORET, *Le Nil et la Civilisation Egyptienne* (Paris 1926) (English translation, *The Nile and Egyptian Civilisation*);

BIBLIOGRAPHY

KEES, *Aegypten* (Munich 1933); ARTHUR CHRISTENSEN, *Die Iranier* (Munich 1933) and *L'Iran sous les Sassanides* (Copenhagen 1944); C. HUART et X. DELAPORTE, *L'Iran Antique* (Paris 1943), L. DELAPORTE, *La Mesopotamie* (Paris 1923); GORDON CHILDE, op. cit. G. FURLANI, *La civilta Babilonese e Assira*, (Roma 1929). On the Islamic countries: A. V. KREMER, *Culturgeschichte des Orients* (Vienna 1875); GAUDFROY-DEMOMBYNES, *Le Monde Musulman* (Paris 1931); A. MEZ, *Die Renaissance des Islams*; (Heidelberg 1922); A. H. LYBYER, *The Government of the Ottoman Empire* (Cambridge, Mass. 1913); GIBB and BOWEN, *Islamic Society and the West* (London 1950); A. BONNE, *State and Economics in the Middle East* (London 1948); J. WEULERSSE, *Paysans de Syrie et du Proche-Orient* (Paris 1946); W. HINZ, *Iran* (Leipzig 1936); FERNAND BRAUDEL's magnificent *La Mediterranée* throws new light on many regions and times (Paris 1949); R. BRUNSCHVIG's *La Berberie orientale sous les Hafsides* (2 vols., Paris 1940-7) is by far the most adequate historical description of an Islamic state. The most penetrating sketches (but unfortunately only sketches) of Islamic social history are contained in C. H. BECKER's *Islamstudien* (2 vols., Leipzig 1924-32), and *Beitr.z.Geschichte Aegyptens* (2 vols., Strassburg 1902-3). *Histoire du Maroc* by HENRI TERASSE is undoubtedly the best general history of an Islamic country yet written. Extremely useful are *Die Goldene Horde* (Leipzig 1943) and *Die Mongolen in Iran* (Leipzig 1938) by BERTOLD SPULER.

On East Asia in general: E. EICKSTEDT, *Rassendynamik von Ostasien* (Berlin 1944). On Central Asia: RENE GROUSSET, *L'Empire des Steppes* (Paris 1939); W. BARTHOLD, *Histoire des Turcs d'Asie Centrale* (Paris 1945); B. VLADIMIRTSOV, *Le Regime Social des Mongols* (Paris 1948).

On India: MASSON-OURSEL, *L'Inde Antique* (Paris 1933) (English translation, *Ancient India*); H. G. RAWLINSON, *India* (London 1943); R. C. MAJUMDAR, *An Advanced History of India* (London 1946); H. GOETZ, *Die Epochen der indischen Kultur* (Leipzig 1929); G. S. CHURYE, *Caste and Race in India* (London 1932); J. H. HUTTON, *Caste in India* (Cambridge 1946); N. K. SIDHANTA, *The Heroic Age in India* (London 1929); R. MOOKERJI, *Local Government in Ancient India* (Oxford 1919).

On China: W. EBERHARD, *Chinas Geschichte*—by far the best—(Bern 1948) (English translation, *History of China*, London 1950); H. WILHELM, *Gesellschaft und Staat in China*—also excellent (Peking 1944); R. GROUSSET, *Histoire de la Chine* (Paris 1942); O. FRANKE, *Geschichte des chinesichen Reiches* (4 vols., Berlin 1930-48); M. GRANET, *La Civilisation Chinoise* (Paris 1929) (English translation, *Chinese Civilisation*); H. G. CREEL, *The Birth of China* (London 1936); R. WILHELM, *A Short History of Chinese Civilization* (London 1929); C. M. WILBUR, *Slavery in China during the Former Han Dynasty* (Chicago 1943); CH'AO TING CHI, *Key Economic Areas in Chinese History* (London 1936); O. LATTIMORE, *Inner Asian*

BIBLIOGRAPHY

Frontiers of China (New York 1940); O. FRANKE, *Staatssozialistische Versuche in China* (Berlin 1931); W. EBERHARD, *Das Toba Reich Nord Chinas* (Leiden 1949); F. MICHAEL, *The Origin of Manchu Rule in China* (Baltimore 1942); WITTFOGEL and FENG, *History of Chinese Society—Liao* (New York 1949).

On Japan: G. B. SANSOM (*Japan*) (2nd ed., London 1946); O. NACHOD, *Geschichte von Japan* (3 vols., Leipzig 1908–30); M. YOSHITOMI, *Etude sur l'Histoire Economique de l'ancien Japon* (Paris 1927); E. HONJO, *The Social and Economic History of Japan* (Kyoto 1935); TAKAO TSUECHIYA, *An Economic History of Japan* (Tokyo 1937).

On South-East Asia: J. NORDER, *Inleiding tot de Oude Geschiedenis van den Indischen Archipel* (The Hague 1948); G. COEDES, *Les Etats Hinduises d'Indochine et d'Indonesie* (Paris 1948); H. G. QUARITCH WALES, *Ancient Siamese Government and Administration* (London 1934).

On Latin America: R. A. HUMPHREYS, *The Evolution of Modern Latin America* (Oxford 1946); H. M. OTS CAPDEQUI, *El Estado Espagnol en las Indias* (Mexico 1941); J. INGENIEROS, *Sociologia Argentina* (Buenos Aires 1946); G. FREYRE, *The Masters and the Slaves* (New York 1946).

On Ancient American Civilizations: S. MORLEY, *The Ancient Maya* (Stanford 1946); G. C. VAILLIANT, *Aztecs of Mexico* (Garden City 1941); L. BAUDIN, *L'Empire Socialiste des Inka* (Paris 1928); H. CUNOW, *Geschichte und Kultur des Inkareiches* (Amsterdam 1937); J. BRAM, *An Analysis of Inca Militarism* (New York 1941).

My data on primitive peoples are derived principally from the following books:

G. BUSCHAN (ed.), *Illustrierte Völkerkunde* (3 vols., Stuttgart 1922–24); H. A. BERNATZIK ed. *Die Grosse Völkerkunde* (3 vols., Leipzig 1939); KAJ BIRKET-SMITH, *Geschichte der Kultur—Eine allgemeine Ethnologie* (Zürich 1946); H. BAUMANN, R. THURNWALD and D. WESTERMANN, *Völkerkunde von Afrika* (Essen 1940); M. FORTES (ed.), *African Political Systems* (London 1940); P. WERDER, *Staatsgefüge in Westafrika* (Stuttgart 1938); S. F. NADEL, *A Black Byzantium* (London 1942); I. SCHAPERA, (ed.), *The Bantu-speaking Tribes of South Africa* (London 1937); G. P. MURDOCK, *Our Primitive Contemporaries* (New York 1934); R. BENEDICT, *Patterns of Culture* (London 1935); C. D. FORDE, *Habitat, Economy and Society* (London 1934); M. J. HERSKOVITS, *The Economic Life of Primitive Peoples* (New York 1940); C. G. and B. Z. SELIGMAN, *Pagan Tribes of the Nilotic Sudan* (London 1932); B. MALINOWSKI, *Argonauts of the Western Pacific* (London 1922); N. DE CLEENE, *Inleiding tot de Congoleesche Volkenkunde* (Antwerp 1943); H. LABOURET, *Histoire des Noirs d'Afrique*; R. H. LOWIE, *Primitive Society* (London 1929), and *Social Organization* (London 1950).

J. STEWART (ed.), *Handbook of South American Indians* (5 vols., Washington 1946–9)—the best ethnographic survey in existence. J. C. VAN

BIBLIOGRAPHY

EERDE (ed.), *De volken van Nederlandsch Indie*, (2 vols., Amsterdam 1920); R. WILLIAMSON, *Social and Political Systems of Central Polynesia* (3 vols., Cambridge 1924). ROBERT MONTAGNE, *Les Berberes et le Mahzen* (Paris 1930); F. C. COLE, *The Peoples of Malaysia* (New York 1945); M. PERHAM, *The Government of Ethiopia* (London 1948).

The fundamental work of OTTO HINTZE, *Staat und Verfassung* (Leipzig 1941), which is a collection of essays and speeches, some dating from the beginning of the present century, reached me when the manuscript was almost ready, but it still helped me to elucidate a few points.

The following books should also be mentioned: FRANZ ALTHEIM, *Weltgeschichte Asiens im Griechischen Zeitalter*, (2 vols., Halle 1948); *Niedergang der alten Welt* (2 vols., Frankfurt am Main 1952).

B. SPULER, *Iran in fruehislamischen Zeit* (Wiesbaden 1952).

E. BIKERMAN, *Les Institutions Seleucides* (Paris 1938).

Glossary of Neologisms

BIATAXY, degree to which distribution of ophelimities in a society is determined by violence or the threat thereof.

BOOKAY, a stratum of warriors, dominating a society.

BOOKAYAN SOCIETY, a society dominated by a bookay.

HOMOIC MILITARY ORGANIZATION, the type of military organization characterized by low M.P.R., low subordination and high cohesion.

INTERSTRATIC MOBILITY, movement of individuals and groups between social strata.

MASAIC MILITARY ORGANIZATION, the type of military organization characterized by high M.P.R., low subordination and high cohesion.

MILITARY PARTICIPATION RATIO, the ratio of militarily utilized individuals to the total able-bodied population.

MONOHIERARCHIC SOCIETY, a society where all hierarchies are integrated.

MORTAZIC MILITARY ORGANIZATION, the type of military organization characterized by low M.P.R., high subordination and high cohesion.

M.P.R., military participation ratio.

NEFERIC MILITARY ORGANIZATION, the type of military organization characterized by high M.P.R., high subordination and high cohesion.

OPHELIMITY, anything which human beings universally desire for themselves.

PARASITIC APPROPRIATION OF SURPLUS, appropriation of surplus produce by individuals not participating in production in any capacity whatsoever; surplus being defined as commodities produced in excess of mere subsistence requirements of producers.

PHERIC DISTANCE, distance measured by the time necessary for covering it.

PLURAL SOCIETY, a society composed of different ethnic groups.

POLEMITY, the ratio of energy devoted to warfare to the total energy available to a society.

POLYHIERARCHIC SOCIETY, a society where several independent hierarchies coexist.

PRETORIANISM, unconstitutional rule of rebellious soldiers.

RITTERIAN MILITARY ORGANIZATION, the type of military organization characterized by low M.P.R., low cohesion and low subordination.

TALLENIC MILITARY ORGANIZATION, the type of military organization characterized by high M.P.R., low subordination and low cohesion.

TRANSCULTURATION, a radical transformation of a people's culture under the impact of another culture.

Index

INDEX

INDEX

237

INDEX